# *In Camp and Cabin*

*Mining Life and Adventure in California, During 1850 and Later*

*by Rev. John Steele*

**with an introduction by Roger Chambers**

# *COVER CREDITS*

**Front Cover**
*Wenderoth Miners Cabin* by Frederick August Wenderoth and Charles Nahls
From the *Bancroft Library, University of California*
[Public Domain],
via Wikimedia Commons

**Back Cover -**
*DETAIL OF FRONT DOOR - Vincent Knorr Cabin, Bettles, Yukon-Koyukuk Census Area*
by *National Park Service*
[Public Domain],
via Wikimedia Commons

**Back Cover Quote -**
*What's the story behind the phrase "There's gold in them thar hills?"*
*True West Magazine* article
Quote from the novel *The American Claimant* by Mark Twain (1892)
https://truewestmagazine.com/whats-the-story-behind-the-phrase-theres-gold-in-them-thar-hills/

**PLEASE NOTE :**
As with all reprinted books of this age that are intended to perfectly reproduce the original edition, considerable pains and effort had to be undertaken to correct fading and sometimes outright damage to existing proofs of this title. At times, this task can be quite monumental, requiring an almost total rebuilding of some pages from digital proofs of multiple copies. Despite this, imperfections still sometimes exist in the final proof and may detract slightly from the visual appearance of the text.

# Self Reliance Books

# introduction

Here at ***Self-Reliance Books*** we are dedicated to bringing you the best in *dusty-old-book-knowledge* to help you in your quest for self-sufficiency.

We're so pleased to bring you another fantastic old title – this time, a nostalgic little book about the life of a Prospector.

**"There's *GOLD* in them thar hills!"** - an old and iconic phrase, popularized by Mark Twain's 1892 novel ***The American Claimant***.

This special edition of ***In Camp and Cabin : Mining Life and Adventure in California, During 1850, and Later*** was written by Rev. John Steele, and first published in 1901, making well over a century old.

The book is a memoir on life back in the old days of the California Gold Rush, and records daily life and adventures in the Goldfields from 1850 onward.

This short, fast read is an absolute must-read for all those interested in prospecting in California, and everyone enthusiastic about the historical context of mining in the region.

~ Roger Chambers

*State of Jefferson, June 2018*

# IN CAMP AND CABIN.

## Mining Life and Adventure in California During 1850, and Later.

BY REV. JOHN STEELE.

### INTRODUCTION.

The following pages are not fiction; but rather confirm the aphorism that "Facts are stranger than fiction," even in common life.

They embrace the writer's experiences and observations in California for about three years, as recorded in his daily journal, beginning in September 1850.

Some of these incidents have gone into history; many will be remembered by surviving miners; and children of the early pioneers, will recall the stories of their father's life in the mines.

Returning to Wisconsin, the author spent some time in study, and was engaged in teaching, in south west Missouri, when the Civil War began; joined the Union army, and at the close of the war, became a minister in the Methodist Episcopal church; and is now a member of the West Wisconsin conference. This journal, written without thought of publication, had been laid aside through all the busy intervening years. Recently, having occasion to refer to it, the author was impressed with the fact, that here was faithfully delineated the every day life and experience of the average miner; and under conditions which only California, in that early day, could furnish.

Here are the various incidents, just as they happened; ludicrous, solemn' serious, tragic, inexpressibly sad, but always interesting.

In Camp and Cabin, is the sequel to "Across the Plains in 1850," and as as that describes life on the Plains in an early day, so this presents daily life in California's most interesting period.

### CHAPTER I.

*In the Gold Mines.—Extremes Meet.—Financial Embarrassment.—Finances Low and Provisions High.—Teams not Marketable.—Hopeful Efforts.—Invest all my Capital.—Rather Despondent.—Getting Points in Mines and Mining.—Renew my Efforts.—First Practical Lesson a Success.—Drifting.—Under ground.—A New Mine.—Poisoned.*

On Monday, September 23rd, 1850, after a journey of over six months our little company reached Nevada City in the gold mines of California, and on the morning of that day, we were all together for the last time.

In the trials of our long journey across the plains, we became well acquainted; mutual kindness and help had taught us to respect each other, and filled our hearts with grateful memories, to be cherished through life. Although their personnel is given in "Across the Plains in 1850," they will appear again in these pages, and I therefore insert their names. Those from Lee county, Iowa; John L. Young, George Matlock, Abraham Hughes, Isaiah J. Hughes, Robert McCord, John Donnelly, Thomas Hunt, Anderson Tade and Drury Farley.

From Iowa county, Wisconsin; William E. Shimmans, Henry Callanan, John Callanan, William Kingsbury, Burton Wait and Thomas Dowson.

From Columbia county, Wisconsin; John Steele. Fifteen men and the boy who kept the journal of the journey recently published under the title of "Across the Plains in 1850."

We were not only wearied with our trip across the plains, but the little money in our possession when we encountered the traders, who had gone out on the trail to speculate with the incoming immigrants, had been paid for supplies of food, in order to prevent actual starvation, until we found ourselves nearly penniless, in a land where bread and meat sold for a dollar a pound.

Our oxen, after their long journey, were not marketable, and were sent to a ranch in the Sacramento valley where it would take months before their skeletons could acquire the requisite flesh to fit them for beef. In the meantime we were sustained by the hope of making fortunes in the gold mines.

But, without tools, or the means to buy, how could we begin? Of course we must find employment, and expect our employers to furnish tools.

Having ascertained that wages, for those who worked in the drifts on Coyote hill, was sixteen dollars a day, we felt happy at the prospect. Gold seemed to be abundant everywhere except in our pockets, and we had faith to believe that they would soon be replenished.

In the early morning of Tuesday, September 24th, my mess, consisting of John L. Young, George Matlock, Abraham Hughes, Isaiah J. Hughes, Robert McCord and John Donnelly broke up, and we started out to begin business in earnest, and all that forenoon I hurried from mine to mine, hopefully inquiring for work.

Sometimes I was asked, "Got any tools?" or "What's your wages?"

Replying, I would say, "I have no tools, just arrived, will work cheap until I get acquainted."

Occasionally there was a little hesitation, but generally an indifferent, "Guess we don't need you."

About noon I went to a restaurant and bakery, hoping to find something to satisfy the demands of an increasing appetite. Taking up a diminutive loaf, I inquired the price.

"Fifty cents," was the reply.

Searching my pockets I found thirty-five cents was my entire cash capital.

"All right," said the baker, "take the loaf, fifteen cents is nothing in California."

That little loaf, only a fair sized biscuit, scarcely enough for a single meal, was all that stood between me and starvation.

Eating about half, and going down to "Rodger William's spring for a drink, (old settlers of Nevada will remember that spring,) I pursued my search for work, but even the eleventh hour passed, and no man had hired me. Night came on: perhaps it was hunger which made me despondent.

Visiting several camps of mine owners, as the workmen came in and were preparing supper, in conversation I learned that there was a special dread lest the "rainy season" might set in, and the deep mines either cave in, or fill with water, causing their abandonment for the season.

I therefore resolved to seek employment in the deep mines, and by diligent inquiry learned enough of the methods of deep mining, drifting, timbering, etc., to intelligently carry on such work.

It was nearly midnight, when, under the shelter of my little tent, refreshing sleep gently stole away my sense of fatigue and hunger.

Awaking with the dawn I was soon on the ridge among the deep mines. In anticipation of immediate work, and probably descending into some damp, cool shaft, I put on my wamus of striped bed-ticking, such as was then worn in the lead mines of Wisconsin, and which I had brought from there.

The few mines employing two sets of hands were in active operation the others silent and still; but within half an hour every windlass was turning, men were descending the shafts, tubs and boxes of gravel were lifted by the long lines.

Hearing a man say thal he had just finished his shaft, and was ready to begin d·ifting, I appiied to him for work. My boyish appearance was not assuring, and my sun burnt face told that I had just arrived at the mines. He eyed me for a moment and inquired, "What do you want a day?"

"Just what you think I'm worth, in fact, I wouldn't mind working for my board until I get acquainted."

He simply said, "No," and went on counting a pile of blocks for timbering.

Removing a few steps, and looking for a place where help seemed to be needed, I saw a slender, sickly looking man, one whom I could have handled with ease, approach the one I had just left, with the inquiry, "Do you want a drifter?"

He looked at the inquirer a moment and asked, "What do you want a day?"

"Sixteen dollars."

"Ain't that pretty high?"

"All right if your mine won't pay it," he said, turning away.

"Hold on, I want you right now, come along."

Here was a lesson for me, and I resolved to profit by it.

Approaching another shaft, which indicated readiness to begin drifting, I inquired, "Do you want a drifter?"

"Yes, what's your wages?"

"Sixteen dollars a day."

"That seems pretty steep."

"It's for you to say."

"Where are you from?"

"Wisconsin."

Just then a stranger to us both chimed in, "When you see a boy from Wisconsin wearin' them togs, he'll do in the mines anywhere."

"All right," said the mine owner, "go to work, and we'll see how it pays."

Stepping into a box about two feet square, its stout rope bail hooked to the windlass line; with a small pick and spade in my hand I was lowered about one hundred feet, to the bottom of the shaft, which was a round hole nearly four feet in diameter.

From the surface of the ground, the first forty feet was through rather loose gravel; then about twenty feet of bluish clay, below which lay very solid gravel, resting on a bed of granite rock. While there was more or less gold scattered, in fine particles, through all the gravel strata, the real "pay dirt" was embraced in a dark colored stratum next to the bed rock, and something over a foot in depth.

The object was to remove this "pay dirt" as quickly as possible, filling it into the boxes and sending it up the shaft. In the meantime carefully separating the "non-pay dirt," and send out as little of it as could be done in order to get at and remove the gold bearing gravel.

Working with all possible diligence, by noon I had introduced and keyed up my first timbers. These timbers were four feet long and about one foot square, split from the large pines which grew on the hill. The posts being set firmly on the bed rock, a little over two feet apart, with a beam extending across the top of the two posts, making a kind of doorway from the shaft, into an excavation which I had already dug to the distance of six feet.

Probably I never did a better half day's

work in California. Though without practical experience in mining, yet by acquaintance with miners in the lead region of Wisconsin, I had learned the importance of properly securing the bottom of a shaft, and so fitting the timber ends to each other that it would be impossible for them to give way.

The miners at Nevada called it, "Squaring the circle."

The shaft being circular, and the earth around it made to rest on a square of timbers, supported by posts, permitting openings on four sides, so that all the gravel underneath could be removed, and the entire structure remain firm, was a problem which involved a practical application of mathematics; but when this was secure, the other timbers were easily placed.

But with anxiety over my work, lest I might make a mistake, and the gnawing of hunger, having gone to work without breakfast, noon found me nearly exhausted; and hinting that it might take some time to prepare dinner, I was glad to dine with my employer, and, in payment, add an extra hour to my day's work.

In the evening, anticipating my need, he gave me an ounce of fine gold, worth sixteen dollars, and, going to the city, I laid in a supply of provisions, prepared my supper, and the next morning took breakfast before going to work.

By Saturday evening, September 27th, with the timbers securely set, test drifts had been run on the bed rock, proving the mine to be very rich. Many of the tubs of gravel, with a capacity of less than half a barrel, contained over one thousand dollars worth of gold.

The next Monday three other hands were set at work, and it required only about three weeks to complete the job, and finish my first effort at gold mining. Mr. Anson Jones, my employer, must have realized a large sum from his mine, how much I do not know, but he seemed well satisfied, and cheerfully paid me up.

In the meantime, having found a vacant place, on another part of the hill, I laid claim to it, and had been doing only enough work to hold it until a thorough test could be made. To this I now gave attention. Two young men, Jim Hayes and Ed. Ogden, who had taken claims adjoining mine, wanted me to buy them out, and, as they asked but little more than the value of their work, I did so.

The tide of immigration had brought down wages to nine dollars a day; so making a windlass and tub, buying a lifting line, and hiring Jim Hayes, I continued sinking the shaft which was already about ten feet deep. We found a trace of gold in the gravel nearly all the way down, and in less than forty feet struck the bed rock.

The "pay dirt" was about two feet deep, and in a little pocket in the bed rock, I found over twenty-five dollars, in fine gold in one pan of gravel. This was very encouraging, and, as the dreaded rains might set in at any time, I resolved to wash only as much gold as would pay expenses, and bring all the pay gravel to the surface, as soon as possible.

Cutting a large pine, near by, and preparing timbers, we were ready for drifting. Employing a man named Mack, (I never learned his full name,) to run the windlass, Jim and I worked below.

While thus engaged, Mack was asked to bring some water from Rodger William's spring.

On his return he lowered it to me, and taking a drink, I passed a cupful to my companion, but just as he began to drink snatched it from him, for I was sure there was something wrong with the water. It was quite clear, with no unpleasant taste, but made me feel very badly.

Hayes had taken very little, but both of us were compelled to leave the mine, and within an hour, it was evident that we would not be able to resume work that afternoon.

When questioned, Mack said he went first to Rodger William's spring, but there were so many people, he would be delayed in waiting his turn, and the constant dipping kept the water muddy, so he went across the creek, and got the water at what seemed to be just as good a spring.

The place was said to abound in cinnabar, but whatever the water held in solution, acted with bad effect on the stomach and bowels. I therefore settled up with the men, arranged for resuming work in the morning, but never entered the mine again.

## CHAPTER II.

*Sick and Alone —Music of the Pines.—Dr. Callanan.—Kind Friends and Good Care.—Sell my Claims.—Walks about Nevada City. —Interrupted Song — Caught in a Shooting Affair.—Oleomargarin.—How Mr. Daniels Failed, and How Mr. Phillips Succeeded.—The Unexpected Happens.*

Suffering from the poisonous effects of the water, I went to my tent and lay down. On the plains we used bacon as an antidote for poisonous water and alkali, so I prepared some, and with hot coffee and bread completed my supper. It made me feel rather worse, so I slept but little during the night, while the wind, through the drooping pine boughs above my tent, kept up a constant dirge-like moan, adding to the feeling of loneliness and depression.

Morning found me too much exhausted to leave the tent, but as the door was on the lower side, I slid down, and placing my head on the split half of a small pine log, which formed a kind of door step, drew aside the cloth and looked out, hoping to see some human being who might be induced to bring me a doctor.

Hours passed; perhaps I sometimes slept, but that dirge-like song of the pine boughs never ceased. At times it was plaintive, sad, depressing; and again, like the full, rich tones of the organ, wonderfully inspiring, just as my own sensitive nerves responded to the strain.

At last a familiar form appeared among the trees, coming up the slope. Raising my hand and motioning to him, he observed the signal, and turned toward the tent: it was Dr. Callanan, the one of all others I most desired to see.

After a cordial greeting, a hurried examination, and fixing me as comfortable as possible on my pallet, he went to the city for medicine, which was duly administered; leaving a dose to be taken "about dark;" promising to see me in the morning, departed with a cheerful "Good day."

The hopeful inspiration of being under the care of a skilled physician, cheered me greatly, and the afternoon and night passed, with only an occasional consciousness of burning thirst. For several days, a confused recollection of the presence of the Doctor and others was all that lingered in my mind. But I had been well cared for; the Doctor had been very attentive, and at his suggestion a Mr. Sexton, an excellent nurse, devoted half his time to my care.

In my time of need, I was fortunate in having plenty of good friends; but before I was able to take care of myself, my expenses had exhausted all my earnings. If I continued to live, more money was necessary, and therefore my claims were offered for sale.

Some looked at them and reported, "They are off the range," others said, "They are too shallow." At last a German, after making a thorough test, offered me two hundred dollars for the entire outfit, claims and tools.

It was the best I could do; so he weighed out the gold, and I gave him a bill of sale.

The claims proved immensely rich; and a week afterward, when I began to walk out, meeting him on his way from the creek, where he had been washing gravel from the mine, he showed me his day's work; a common wooden water bucket, more than half full of fine gold. And this was only one day's washing. I could not estimate its value, but its weight seemed to be all the bucket could support.

When sufficiently recovered to walk, I naturally strolled around the city. There were a number of good stores, meat shops, bakeries, blacksmith shops, etc., but the gambling saloons were the terror of the town. Their rooms were spacious, supplied with music, adorned with mirrors, pictures, and every device to attract the young, and induce them to gamble and drink.

Boys and young men, from respectable homes, from quiet villages and country places in "the states," here spent their evenings, and formed associations and habits which wrought their ruin. Here too, men crazed with drink, and maddened by losses, either killed themselves or others.

In those days, scarcely a night passed that men were not killed, in or about the city. There seemed to be no organized government; or if such existed, people were too busy with their own affairs and interests to give attention to the execution of law; but so far as possible, each one tried to protect himself.

Fortunately, none of my associates were inclined to visit the saloons, though it must be confessed that there was such a witchery in the music, instrumental and vocal, that the masses were attracted and entranced, and in passing I found it difficult to resist the temptation to go in and listen.

One night, however, when trying to see a man in relation to some business matters, I was directed to a large saloon. From the door, through the smoky air, I saw him in a distant part of the room Pausing to see how best to reach him through the throng, a man, with a large revolver in his hand, passed me, quietly pushing his way through the crowd; so I followed in his wake.

As we neared the center of the room, I noticed that the hum of voices, representing all possible tones, was gradually hushed by a song, which, from an elevated platform, a young man was singing to the accompaniment of a violin. For a moment the rudest became quiet. All were intently listening to the thrilling strains of Burns' Highland Mary, and leaning to catch every word, as with softest pathos he sang,

"O pale, pale now the rosy lips,
I oft have kissed sae fondly,
And closed forever the spark'ling glance
That dwelt on me sae kindly."

Just then the man I had followed flourished his pistol, and breaking the breathless hush of sweet melody by a hoarse volley of bitter oaths, commenced firing at a man who faced us, and was, apparently, trying to reach the door.

Instantly all was uproar, and rush for the door. The victim of the assault vainly tried to reach it, but fell a little inside. The shots into the crowd were effective; three were killed, and several wounded; and then the assailant sprang over the bar, passed out at a back door, and made his escape.

I learned afterward that two of the men killed had no part in the quarrel, simply, by accident, came in range of the assassin's pistol, and lost their lives. The whole affair impressed me with the folly and danger of being a mere spectator at such a place: a lesson of practical value to me.

During convalescence I became acquainted with Mr. Daniels, from Baltimore, who kept a small supply of groceries in a tent, and, as his place was convenient, I did most of my trading with him. He manufactured butter from tallow and lard, and it looked and tasted so much like real butter, that, without comparing it with the genuine article, which long since had been only a memory, I could not tell the difference. However, he deceived no one, but sold it for just what it was. He never explained the process of its manufacture, and whether he was the originator of oleomargarin I do not know.

One afternoon he introduced me to his friend, Mr. Phillips, and then related the following story. When the California gold excitement reached Baltimore, Daniels was a dealer in general merchandise, and Phillips employed by the month as a porter in his store.

Thinking he saw an opportunity to make an immense fortune, Daniels closed his business in Baltimore, and to the value of $25,000, selected groceries, clothing and hardware, such as he thought suitable for the California market, and shipped them to San Francisco by way of Cape Horn.

Phillips wanted to go to California, so Daniels loaned him the amount necessary to pay his expenses, and on reaching San Francisco, Phillips hired to work by the day in order to earn his passage money, and pay his way to the mines.

Daniels on reaching San Francisco with his goods, sold part to pay storage and transportation for the rest, and at last succeeded in getting what was left to the mines on the South Yuba.

Just then the Nevada mines were discovered, and he found himself deserted by the miners, who flocked to Deer creek, and he was compelled to move his goods to Nevada City.

By this time his capital stock was greatly reduced, and upon making an inventory he found that, even at the high rates charged in the mines, the value of his remaining stock was far less than when he left Baltimore. In fact he was nearly bankrupt, and was now disposing of his limited stock, with the expectation of taking up the pick and shovel, and trying his fortune in the mines. He had not only lost much precious time, but his capital was nearly gone.

In the meantime, Phillips, according to his reckoning, had earned enough to repay Daniels, meet his expenses to the mines, and supply himself with an outfit of tools. He therefore sought a settlement with his employer, for as yet he had drawn but little of his wages, but found it impossible to obtain his pay in cash, and after considerable delay was compelled to take some city lots or nothing.

However, after receiving his deeds, in due and legal form, he started out to find his property; thinking perhaps he might have a place to pitch a tent, but he was utterly disappointed and disgusted to find every lot covered with the waters of the bay, and they seemed to be of little or no value.

Homesick and discouraged he hunted another job, careful now to draw his pay every week, and deploring the hard fortune which kept him away from the mines, for which he had come so far.

While engaged as porter at a hotel a real estate dealer asked him whether he owned certain lots, giving the number and description; and receiving an affirmative answer, replied, "I knew they belonged to a man of your name, but did not think that you were the one; we have been looking them up, and if you want to sell, come to our office, and I think we will buy them."

Some time afterward, a man who had overheard the conversation, said to him, "I think I could afford to give you fifty thousand dollars for those lots."

This set him to making inquiry in regard to prices and buyers. The rapid development of that part of the city had brought them into demand, and he finally sold them for eighty-three thousand dollars.

Now, he had no need to go to the mines, except to repay his benefactor, and hence his trip to Nevada City. But how strangely the fortunes of those men had changed, since leaving Baltimore.

How often it happens thus. The most carefully arranged plans and confidently expected success brings only disappointment, while apparent failures and expected disappointments result in unexpected success. Sometimes it seems enough to induce people to suspend judgment, abandon their plans, and trust to luck; but the wise ones never do so, knowing that "to err is human," judgment is matured, and plans perfected; and yet, there was a proverb common in California: "It is the unexpected that happens."

## CHAPTER III.

*Unfortunate, But Hopeful.—Prospecting.—Tired Out.—Sleep in the Woods.—The Grizzly Bear—Again in the Coyote Diggings.—Mine Caves In.—Kicking the Bucket.—Escape—Resume Work.—Jefferson on South Yuba.—Mining on South Yuba.—Arrange a Trip to the Feather River.—Delay.—Visit Nevada City.—Death of George Matlock.—Visit Messrs. Zachary Bowers and Abijah Davis, of Wisconsin.—Mr. Dinkler.—Compelling a Settlement.*

It would be impossible to describe the depression which for a time came over me as I realized that not only had a fortune slipped from my grasp, but health had also gone with it. However, youth and hope whispered of returning health and boundless opportunities, and therefore as strength permitted, my time was devoted to prospecting.

The hills, ravines and gulches about Nevada were visited only to be found unpromising, occupied or claimed. Still, rich mines on all sides showed how hope to others had ended in fruition, and lured me on. Each night fatigued, but with the dawn hopeful and ready to renew the search.

Returning from one of my prospecting tours and ascending the ridge between Deer Creek and the Yuba, I struck into the immigrant road, and turning toward Nevada intended to pass the night at Cold Spring Cottage, a hotel about fifteen miles from the city.

It was late in the afternoon, and, in the shade of the dense pine and cedar trees, darkness came suddenly, and I groped my way to the hotel only to find it closed.

There seemed no alternative but to trudge on to Nevada; but, weakened by recent sickness, worn with the toil and journey of the day, and also without supper, I found the task too great, and utterly gave out.

So turning a short distance from the road, and creeping under the drooping branches of a low cedar, I lay down to rest and sleep. Sleep did not readily come to my relief. Sometime in the night I heard a team and wagon pass along the road, but it was not going towards the city, and therefore could afford me no aid. At last, though somewhat chilled, I waked from a refreshing sleep to find that day had come.

But had I known what was passing around me during the night, I would have spent it among the topmost branches of the cedar, rather than on the ground, for when resuming my journey I found the broad footprints of a grizzly bear. It came from the opposite side, followed the road for some distance, showing that the tracks were made since the wagon had passed, and finally turned in the direction toward where I lay, and must have come quite near.

How he failed to find me was inexplicable, and I shuddered at the thought of awakening in his strong embrace; but made up my mind never to give a bear another such chance.

Afterward, while reading the Bible, I came to these words, which recalled the incident and impressed me deeply: Psalm iv., 8. "*I will both lay me down in peace and sleep, for thou, Lord, only makest me to dwell in safety.*"

Truly, God's providence is a better protection than human wisdom, strength and skill; for when we have exhausted all these we are still safe under his care.

Reaching Nevada, again the specter of starvation began to haunt me, but I was, at least, thankful for improved health and having been saved from the grizzly's teeth.

About this time I accepted Fred. Dinkler's offer of nine dollars a day to work in his mine on Coyote Hill. It was some sixty feet deep, originally embraced four rods square, but was nearly half worked out. The timbers, especially around the shaft, were in a very unsafe condition.

There were two sets of hands; the one to which I belonged, going to work at noon, working twelve hours, was relieved at midnight.

It was November, but, as yet, little rain had fallen, and Fred hoped to have the mine worked out before it was flooded, or made unsafe by the rains, so the work was pushed with all diligence.

One night about ten o'clock, we heard a crash in the shaft, and, immediately, the timbers began to settle and break. There were five of us in the mine; four in the drifts, and one who dragged the buckets to the shaft, unhitched the empty one, and hitched the full one to the windlass line.

Part of the shaft had caved in, and, to all appearance, we were about to be buried in a deep grave. The shaft was so filled that the bucket could only be lowered to within about four feet of the bottom; and as we listened, we could hear an occasional boulder, coming down, strike against it. Just then some one provoked a smile by saying, "Boys, that sounds like kicking the bucket."

But when our candles became dim, indicating that the air was shut off, that slang allusion to death had a terrible meaning. However, the man at the shaft worked his way up, opened a passage into the shaft, got into the bucket, and was safely raised to the top. Again the bucket was carefully lowered, and so, one by one, all at last escaped.

When I went up, I carried my candle, in order to examine the shaft, which seemed firm until near the top, where the gravel was loose; and a considerable bank had caved off. The boulders continued to fall, with an occasional thud; but, fortunately, none fell while any of us were on our way up. Of course, the midnight relay could not go to work, and most of the workmen never again entered the mine.

At noon the next day, after examining the shaft, and removing the boulders that were likely to fall, three of us went down, sent up the gravel from the bottom of the shaft, and entered the drifts. The posts all leaned a little toward a worked out and abandoned mine, which had caved in; and the immense weight of earth was indicated by the heavy timbers being bent over the posts in the form of ox yokes; and the great mass of earth overhead had settled eight or ten inches.

After placing posts under the broken timbers, and adding a few extra beams, we resumed digging, and although conscious of danger, four of us continued the work until the "pay dirt" was all removed.

Discouraged with the prospect at Nevada, in company with Robert McCord and Drury Farley, I went to the south fork of the Yuba river. From the "Sugar Loaf" hill above Nevada, we followed a trail among large pines and cedar trees, on a broad upland between Deer creek and the Yuba, striking into the immigrant road near Cold Spring Cottage. Taking the road eastward until we found a trail, which we had been told would lead us to Jefferson on the Yuba, and following down a steep, winding spur, in about three miles reached a trading post on the river bank, with a few deserted cabins in sight.

This was Jefferson, about twenty-five miles from Nevada; the stream, shut in by almost perpendicular mountains, was about four rods wide and two feet deep; but being a series of eddies and cascades it was difficult to estimate its size.

Considerable mining had been done along the river, but when the Nevada mines were discovered, these were abandoned; and merchants, who, at great expense had brought their goods, were compelled to pack them to other places.

About a mile above Jefferson, on the

river bank we found a narrow bar, which had been overlooked by the prospector. Digging down to the slate bed rock, and washing a pan of gravel, we found about a dollar's worth of gold in fine, bright scales.

This was encouraging; and repairing a cast-away rocker, we went to work in earnest; made a thorough test of the bar, and in two days, with our defective rocker, succeeded in taking out over a hundred dollar's worth of gold.

Next we arranged for putting in a long-tom and sluice. We found sufficient lumber in the abandoned mines, but it required a journey to Nevada to obtain material for hose. This was simply strips of stout drilling, sewed in the form of a pipe by which the water was conveyed from the river into our sluice.

While our work on the bar lasted it paid well, but in three weeks the narrow strip of "pay dirt" had all been washed, and we moved to another place.

In the meantime I became interested in a company, organized at Nevada, for the purpose of taking provisions and mining implements to the north fork of Feather river, a place said to be very rich in gold.

The company numbered twenty, and forty mules were loaded, mostly with provisions; and eighteen of the company, and the four men who owned the mules, started in November. One of our number, a young man named John Donnelley being ill, and, as I was profitably employed, concluded to wait for him.

Saturday, Dec. 14th, 1850. This afternoon Drury Farley was informed by a messenger, that his friend, Anderson Tade, near Nevada, and with whom I became acquainted on the plains, was very ill, and desired him to come. As I expected soon to start for Feather river, and having business to arrange at Nevada, the next morning McCord and I accompanied Farley, arriving at the city late in the afternoon.

I now learned of the death, a few days before, of my very dear friend, George Matlock, with whom I had crossed the plains, and who had been instrumental in saving my life. A man whose Christian character was more than a profession: sweet in spirit, self-denying, ready to make sacrifice for the sake of others. A leader in every difficult and dangerous enterprise, and one whose prudent, firm, intelligent courage insured success. My own father could not have treated me with greater kindness than I received from him. But alas! alas! the inexpressible loss and sorrow of his widow and family when the story of his death reaches them in their far distant Iowa home.

Tuesday, Dec. 17th, 1850. This day I visited Messrs. Zachary Bowers and Abijah Davis, acquaintances from Wisconsin, whom I found engaged in mining on Deer creek, about two and a half miles below Nevada.

When I quit work for Mr. Dinkler he paid me only a small part of my wages, saying, when he had time to wash the gravel, within a few days, he would pay the rest. Weeks had passed, and now, after three days failure to find him, I began to suspect that he was trying to evade me. When told he was at his cabin, and going there, I always found that he had just gone. Again and again, when a time was set to meet him, where his workmen said he expected to be, he failed to appear. It was reported that though he had taken large quantities of gold from his mines, he would never pay a dollar if he could help it. At last, learning that two men expected to begin work for him in the morning, I went to the place at an early hour, and sitting among some mine timbers awaited his approach.

He was promptly on hand, and I, just as promptly, claimed his attention. He was surprised, but soon regained his presence of mind, and said, "I knew you were in the city, and if I could pay you, would have hunted you up." "But, said

I, "any of these grocerymen will lend you, for a few days, the small amount you owe me."

"O, no, no, no, people don't lend money without security."

"Of course, but you can give security; step into this store and we can settle this matter in a few minutes."

"Wait till I get my men at work."

"Yes, I'll wait."

The men were already at work, lowering timbers into the mine, so, with reluctance he accompanied me into the store. Explaining to the merchant the circumstances; that I was about to leave, and would he not be so kind and obliging as to loan the money to Mr. Dinkler for a few days, and, of course, he could give ample security.

"Certainly," said the merchant, I *could* advance the money, but I believe he has it, and if he won't pay you without trouble, he would not pay me."

This seemed to settle the matter, and a look of satisfaction came over Dinkler's face, as he turned to go out. There was still another resort, and I resolved to frighten him into the payment.

I sprang before him into the door; presenting a pistol, with a loud voice ordered him to "Stop! Now sir, I'm going away this morning, but this matter must be settled first; you can pay it now, or never have another chance."

His voice trembled as he shouted, "Don't, don't shoot!" And springing to the counter, upon which stood scales for weighing gold, he drew from his pocket a large buckskin purse of the shining metal, weighed out the amount of my claim, and handed it to me.

The high words and flourish of the pistol attracted attention, and men on their way to work crowded into the store. The merchant explained the matter; and when Mr. Dinkler was about to leave, several blocked his way, saying, "No, Fred, it's your treat; you intended to cheat that boy out of his wages,—now you shall treat the crowd; set out the cigars."

How many were taken I do not know, but the amount of the "treat" must have been nearly as much as he had owed me. However, he silently weighed out the gold, and the crowd dispersed.

## CHAPTER IV.

*Return to Yuba River.—Deep Snow.—Joined by John Donnelly.—Washington.—Canonville.—Poorman's Creek.—Drury Farley and Thomas Hunt.—Death of Anderson Tade.—Sickness Among Miners.—Donnelly and I Start for Feather River. Snow and Ice.—"Wicked Stand in Slippery Places."—Evidence of Murder.*

Wednesday, December 18th, 1850. Returning to the Yuba, McCord and I encountered deep snow on the uplands. For several days there had been heavy rains in the valleys, and, at the same time snow had fallen to a great depth on the mountains.

After a few days, Donnely having recovered from his illness, and ready for the journey to Feather River, joined us at the Yuba; but now that the snow was so deep and soft, we hesitated about starting on our northern trip. However, we improved the time in mining; did a great deal of prospecting along the river; found an occasional rich "pocket," but the general outlook was not promising.

Two or three miles above Jefferson, on the south bank of the river, stood the deserted village of Washington. With a large number of vacant cabins it contained several empty store buildings, and quite a large hotel closed and silent.

A few miles farther up, where the rocks rose perpedicularly on both sides of the river, leaving but a narrow margin, was the hamlet of Canonville, entirely deserted.

Lumber had been sawed, and the river taken from its channel and carried some distance in a flume, thus laying bare the bed, but the enterprise did not pay, and after large expenditure of time and money was given up.

Above this there seemed no chance for mining, as the river ran between perpendicular banks of rock. At Washington there were a few miners, like ourselves, prospecting, and so, in places, along the river. During the past summer a vast amount of labor had been expended in this vicinity, but it evidently failed to pay expenses.

Finding a fair prospect on Poorman's creek, a small tributary of the Yuba from the north, we built a cabin and for a while our mining operations were quite profitable.

Wednesday, January 1st, 1851. This afternoon Drury Farley, and Thomas Hunt, an elder half-brother, arrived from Nevada City, bringing the sad news of the death of Mr. Anderson Tade, who, in the closing hours of the old year, they had laid by the side of his friend and neighbor Mr. George Matlock. His widow and family, whose home is near Fort Madison, Iowa, must receive the painful message that, while hoping to brighten their lives and better their condition, he had traveled thus far from them only to find a grave.

Throughout the mining camps there was much sickness and many deaths, occasioned doubtless in part by scanty and stale provisions, which induced scurvey; also, working in damp places, under ground and in the water. With proper food and shelter, the climate must be healthful; but the conditions under which most miners lived and labored invited disease and death; and it was difficult to better the conditions.

Although we had planned to make the trip to Feather River before snow had fallen on the mountains; and all the company except Donnelly and I had gone; while profitably employed we were willing to wait; but having worked out our mine, and believing that the snow on the uplands had settled and become hard; we resolved to push out for Feather river.

Being joined by nine others, who went as prospectors, we decided to start on Monday, the 27th of January, 1851. But when the morning of our departure dawned, a misty rain made us hesitate until 10 A. M., and then, like a train of packed mules, we filed up the mountain. Besides our blankets, some extra clothing, rifles and amunition; Donnelly and I carried a pick and spade; pan for washing gold, frying pan, tin cups; and with some bread, flour and bacon enough to last two weeks. All were equally well, loaded, some even more heavily.

Following up a very steep, rocky spur, early in the afternoon we came out on the "divide" between the south and middle forks of the Yuba. Here snow was several feet in depth, and softened by the mist which continued, we sank deeply; and weighed down by our heavy burdens, made slow progress.

About dark, finding a grove of large fir trees, and beneath them but little snow, we camped, built large fires, prepared supper, and placing plenty of fir boughs on the ground, over which we spread our blankets, "Lay down to pleasant dreams."

Looking up through the long, drooping branches which canopied our sleeping apartment, we saw that the clouds had cleared away and the stars blinked brightly down.

Preparing breakfast before day, and shouldering our packs, we were away with the dawn. It was our hope that the rain would settle the snow and the frost of the past night make a crust sufficiently hard to bear us up. In this, however, we were disappointed. In places we could walk a few steps on the surface, but generally broke through, and as the snow was very deep, passing over some shrub or hidden branch, we would sink to our shoulders.

Thus we floundered on, and early in the afternoon, down through a chaos of cragged ravines, and about three miles distant, obtained a glimpse of the middle fork of the Yuba, across which lay our route.

In our descent at first we were aided by the snow, but this gradually became less, and being on the north or shady side of the mountain, the snow terminated in a vast sheet of ice, over which the greatest care was required to keep from going down too rapidly. After rolling some distance, Mr. McGee, trying to steady himself by a small pine, remarked, "The Bible says that the wicked stand in slippery places, but I can't and there ain't one in this crowd who can."

Accompanied with a chorus of loud talk, a clatter of camp kettles, tin pans and ovens, the party at last reached the river. Considering the rapidity with which some of us came down, the rocks, crags and projecting roots over which we glided, it was a marvel that there were no broken bones; but all had suffered, more or less, from bruises, scratches and torn clothing.

Crossing the river on a flume, by an Indian trail we ascended the opposite mountain, now on the sunny side, and, in about seven miles from the river, camped at Indian creek.

Wednesday, Jan. 29. Making an early start, we ascended a steep, rocky, treeless spur about two miles, and then entered upon a slope, mantled with large pine, fir and cedar trees and crossed with an occasional cliff.

For awhile our path was quite pleasant, but on the uplands we again encountered snow, and, as we ascended each slope, it became deeper, until only the tree tops appeared above its surface. Fortunately it was hard enough to bear us up, unless we trod in the vicinity of a tree top, when we were liable to go down among the branches.

As we advanced, the great summit ridge of the Sierra Nevada toward the north east towered above forest and cliff, and reminded me of my first lessons in the old English Reader,

"The increasing prospect tires our wandering eyes,
Hills peep o'er hills, and Alps on Alps arise."

Descending the mountain through a magnificent forest of the usual pine, fir and cedar, about dark we camped on the mountain side half a mile from Dowineville, a mining village at the forks of the north Yuba.

Thursday, Jan. 30. The sky being overcast with clouds, and a slight rain falling, made us hesitate about starting, and in the meantime one of our company came into camp saying, "I reckon somebody has struck it rich down there, and covered up their prospect hole so as to hide it."

With picks, shovels and pans, three of us accompanied him to the bottom of a deep, wild glen. Not that we intended to "jump" any one's claim, but as a possible clue to diggings above and below on this side of the river. There was no snow, and on the mossy bank of a rill could be seen the outlines where the ground had been broken; but the turf was so nicely adjusted that few traces were visible.

Spading away the soft earth to the depth of about three feet, we found,—not a gold mine, but that which made us start back with horror,—a blue shirt sleeve on the arm of a corpse. Gently the body was uncovered and raised to the surface; water was brought and, washing away the mire, disclosed the features of a young man, of probably twenty years; about five feet in height; dark brown hair; his only clothing a blue woolen shirt, dark brown pantaloons, and heavy boots.

His pockets were empty and nothing about him to reveal his name. Traces on each side of his head indicating where a bullet had passed through, were the only marks of violence upon his person. Evidently he had been murdered but a few days since and his body concealed in this wild glen.

Tears filled our eyes as we thought of

his untimely fate, and that father, mother, brothers and sisters may lovingly await his return until hope deferred makes the heart sick.

The death sealed lips could not reveal the name of the murderer to men, but there is a Witness who knows all about it, and sometime the criminal will stand at the judgement bar of God.

The remains were taken to Downiville, and without being identified, were buried there. Long afterward when passing, I made diligent inquiry, and learned that no knowledge of his name, friends, or home had been found. To use a phrase common among mountaineers, he had been "rubbed out."

## CHAPTER V.

*Downieville.—Ascending the Mountain.—John Cheny.—Extemporized Culinary Outfit.—Illuminating the Camp.—John Cheny's Camp Fire.—The Yuba Caps.—Crossing the Divide.—Canon Creek.—James Ward's Story.—One Man Frozen to Death and Three Lost.—Downie's Diggings.—Donnelly and I Continue our Journey.—Puma Tracks.—Beyond the Miners.*

In and about Downieville some very rich mines had been discovered, but at this time the place seemed over run with prospectors. While some were making fortunes, and others doing fairly well, a great many without mines or work, were in desperate straits.

Some of our company found acquaintances and concluded to remain. Late in the afternoon the rest of us left the village, and ascending the northern mountain about eight miles, camped in a dense forest; where, shoveling away the snow, and spreading boughs of fir and cedar, on which to lay our blankets, made a comfortable place to sleep.

The older members of our company were greatly fatigued with wading the deep snow, and those who were younger relieved them of part of their burden. But with all the toil and exposure, there is something invigorating in this mountain air, which sharpens the appetite, and promotes health, so that some of our party seem never to tire.

John Cheny, a young man of eighteen, strong and healthy, greatly enjoys this out-door life. This afternoon he carried a fifty pound sack of flour, his blankets, pick, shovel, and gun, and yet, in the steepest and worst part of the journey, he relieved an elderly man of a large roll of blankets.

As for victuals, we have learned to simplify the process of cooking, and perhap regard quantity rather than quality. When the camp fire is built, a mass of snow held near it on a wooden fork, soon becomes like a well filled sponge, and furnishes water for coffee and drinking purposes.

Also in making bread, a little snow put into the mouth of a sack of flour and kneaded carefully, until a stiff dough is formed; then lifting it out, and molding it with the hands until of proper consistency for bread. It is then suspended near the fire, on a bough, with several branches cut and sharpened for the purpose. Turning it occasionally, the cake is soon thoroughly baked.

Sometimes a slice of bacon is suspended on a wooden fork by the fire, and as it fries, the fat is permitted to fall on the bread, thus making it more palatable. It is astonishing how small a culinary outfit is really needed.

Just above our camp stood several large dead pines, probably fire killed, but overgrown, from bottom to top, with long, yellow moss. After dark, setting this on fire, the flame soon streamed far above the woods, making immense torches, and illuminating our camp; and, as the wood was filled with pitch, they continued to burn most of the night.

After all my companions had lain down to sleep; while writing up my journal, I noticed that one of the burning trees was about to fall, and, fearing it might come down upon the camp, watched it until there was evidence that it would fall across the rocky slope above us. Therefore, without waking my companions, I spread down my blanket, and was about to join them on our bough built bunk, when the tree fell, breaking in several pieces on the rocks; one great fiery mass rolling directly over our bed, stopped against the logs which composed our camp fire.

The crash awakened the sleepers, and while they all escaped, there was no time to remove the bedding. However, shaking the coals from our blankets, and changing the boughs to another place, most of them were soon again soundly sleeping.

Meanwhile John Cheny planned to bring two more pieces of the burning tree, and with the three great logs piled together, have a splendid camp fire. With considerable effort he brought the second piece, then prepared skids and aroused the camp to help him bring the third But the others were too tired and sleepy to respond.

Finally McGee advised him to lie down and not disturb them.

"Well," replied Cheny, "we'll not leave here till morning any way, and we might as well be rolling logs as doing nothing else."

This was his characteristic: rest and sleep seemed quite unnecessary; so he persevered, obtained help, and when the three immense logs were properly grouped, and wrapped in roaring flames, he lay down to enjoy his watch fire, and was soon asleep.

Friday, Jan. 31. A severe frost so hardened the snow that we walked on the surface, and taking an early start, were soon above the line of heavy timber, and by noon up a slippery ascent, reached the "Yuba Caps," a mass of perpendicular rocks, which crown the summit of the mountain.

The younger members of the company had an ambition to climb these rocks; this, however, we found to be impracticable at this side, but from their base we obtained a magnificent view of the Sacramento valley and Coast Range.

Around the bottom of these rocks the snow was almost perpendicular, with a surface of hard, smooth ice. There was no alternative but to cut footsteps in the snow, and thus pass around toward the right. As it was very laborious, we took turns, one going forward with the hatchet, and the others following in his footsteps. In places it was so steep as to require handholds as well as footsteps, and some experienced great difficulty in keeping their balance, as they looked down from the dizzy hight, which we estimated at about a mile, where, if one missed his footing, he must fall.

In looking down, it seemed like one unbroken sheet of icy snow; but we knew, a discovery made while coming up the mountain, that it was crossed by several precipices of great hight. Just how far it was around the "Yuba Caps," we could hardly guess; to us it seemed about two miles, and it was a relief, when, crossing the "divide" toward the north, we again reached the timber line.

At Canon creek, in a grove of small fir trees, we made a place to camp, by shoveling away the snow, which was about four feet deep.

Saturday, Feb. 1. The day was bright and cold, and we made good progress over snow of great depth. In a small valley we found a place where wolves had burrowed in the snow, and brought to the surface tufts of hair, which indicated that horses or mules had perished there; but we passed it without special attention. However, the next afternoon, descending from the "divide" to Downie's diggings on Poorman's creek, we found James Ward, one of our company who left Nevada last fall, painfully going about on crutches, and from him learned the secret of the wolf holes.

He informed us that they had a very pleasant trip to the north fork of Feather river, where they pitched their camp, and prepared for a winter's work. He, and four others, then undertook to bring the forty mules back to Nevada City.

Some time in December, while on the divide, they were overtaken by a severe snow storm, and took shelter for the night in the valley above named. By morning they were snowed in, and after remaining two days, and the storm still continuing, feeling sure that the mules must perish, they tried to save themselves by going back to Downie's diggings.

The snow had whitened the trees and rocks, and as it still continued to fall, the outlines of the mountains could not be seen; consequently they lost their way; disagreed as to the proper direction and separated. Ward and one other taking one direction, and the three others the opposite, and, so far as I have been able to learn, were never again heard from.

Ward and his companion struggled through the snow during the day, and at night, climbing down the branches of a fir tree, buried themselves in the snow; and, as they had blankets and provisions, were quite comfortable. But the following day was intensely cold, and the next morning, coming out of their shelter from under the snow, they could not agree as to the direction, and finally they separated. Fortunately, the same day Ward was found by his brother Thomas, who had a mine on Poorman's creek, and was crossing the ridge to Onion valley on snow shoes.

It seemed the merest chance that they should meet, and had James followed the ridge in the direction he was going when found, he would have passed the only camp for many miles, and must have perished.

As it was, his feet were so badly frozen, that for more than a month he had been unable to walk without help.

After considerable search, his companion was found on a rock on the summit of a high ridge, which he had climbed, evidently in the hope of seeing some landmark to guide him on his way; but, overcome with cold, had frozen to death.*

At Downie's diggings, on Poorman's creek, Donnelly and I, reluctantly, parted with those who had accompanied us from Poorman's creek near the south Yuba; both streams, we were told, deriving their names from the same pioneer miner, Mr. Poorman.

Having obtained all the information possible from Mr. Ward, as to our route, and the location of our company, we replenished our stock of provisions, and were ready to pursue the journey

Monday, Feb. 3. From nearly the summit of the "divide" between the Yubas and Feather rivers, we followed down Poorman's creek to its junction with Hopkin's creek, then down this to Nelson's creek; occasionally compelled by impassible canons to leave the valley, and cross high, steep, rocky and icy spurs.

About dark we waded Nelson's creek, and ascending the mountain side camped among thickets of manzanita, which furnished a supply of excellent fuel for our camp fire.

The morning was fair, but in the afternoon clouds gathered, and there fell a slight rain, which made us regret our shelterless condition. But when we lay down to sleep beside our camp fire, the rain had ceased, and the stars twinkled encouragingly above us.

The next day was pleasant and we made good progress; over a high mountain; and toward night descending into a creek valley where we found the tracks of the puma, or California lion; spent some time in hunting it; but from what we afterward learned, respecting the size and ferocious nature of these animals, it was just as well that we failed to find him.

One object in selecting our route and making our journey in this way, was to obtain a general knowledge of the gold mines. We might have selected an easier way, but we wanted to visit the best mining region, and that took us across the spurs of the great Sierra Nevada.

Heretofore we had found plenty of miners at work, and could gather information from them; now, however, we were beyond the usual range of prospectors, and therefore made it a point to examine the bed and banks of all the streams we passed, hoping to find gold in such quantities as to make profitable a return, at some future time.

Not finding even a trace of gold along the creek, night closed our work; and kindling a large camp fire, we cooked supper, and were soon asleep.

## CHAPTER VI.

*Across the Mountain.—Dangerous Descent.—Middle Fork of Feather River.—Search for a Crossing.—Swimming in Ice Water.—Deep Wading.—North Fork of Feather River.—Cold Reception.—Line of Flags—Village of Hostile Indians.—Passing the Village.—Scant Rations.—Fears.—False Alarm.—Intense Cold.—Big Meadows.—Immigrant Road.*

Wednesday, Feb. 5. Our way led over a high, steep, heavily timbered mountain, on which the snow was very deep, and just at sunset we began the descent to the middle fork of Feather river. The deep snow at first rather assisted, but, terminating in ice, the way became more steep, and was crossed by an occasional cliff.

While the light lasted we could mark our way, and slide from tree to tree, and sometimes, by rocks and branches of trees, swing ourselves down the ledges; but clouds overcast the sky, the light faded, and darkness became intense long before we reached the river At times, as we lingered on the brink of some precipice, and tried to rest or plan for the next move, it seemed as though we could neither retain our position, nor go on with safety.

A pack of mountain wolves were on our trail, and their fierce howls, mingled with the deep bass of the river, reverberating from the canon below, and all strangely softened and subdued by the sigh of the pines around us, seemed intoning a dirge, in weird, depressing voices from out the night.

At last we reached the river in safety, somewhat bruised and scratched, but just how we made the trip in the dark, it is doubtful whether either of us could tell. Donnelly had his outfit intact; I had lost my haversack containing my Bible, writing material, journal, and most of my ammunition. As I stumbled on the brink of a precipice it slipped over my head, and had fallen below the cliff.

Kindling a fire on a little sand bar, we soon had the wolves at bay. As our guns were wet, we took off the barrels, placed the breech in the fire, and, when sufficiently dry, they were fired, wiped, and reloaded. Then after supper, replenishing our fire, or rather, from a large pile of drift wood, building two, we lay down on the sand between them, and the wolves serenaded us until we slept.

The next morning while Donnelly prepared breakfast, I went in search of my haversack, which I found lying at the foot of a precipice over a hundred feet high. As I looked up, and thought how near I came to falling over it in the darkness, a cold shudder crept over me.

It was our intention to cross the river at this point, but we had descended into a canon, and the swift current and perpendicular banks warned us not to make the attempt. So we spent the entire day in search of a crossing, and at night camped on the same side, several miles below.

The threatened storm had passed, the sky was clear, but being on the shady side of the great mountain, we scarcely saw the sun, and our way was very icy and dangerous.

The next morning we found a place where the water was not very swift, but, without an axe, were unable to obtain logs large enough to float ourselves across. However, preparing a little raft of dry branches, and placing our guns, packs, and clothing thereon, we pushed it boldly into the deep, clear current, and were soon at the opposite bank. The river was fringed with ice, and cold as was the water, the air seemed even colder, and made our teeth chatter with the chill; but hastily dressing, and shouldering our loads, the exercise of climbing the mountain soon warmed us up.

After making the ascent, and crossing some deep ravines, we found an Indian trail, and as it lay in our course, followed it. After several miles it descended to the head of a small stream, down which we were led, through a broad and beautiful valley, at night-fall camping on its bank.

Saturday, Feb. 8. Early this morning, leaving the stream to our left, we crossed some heavily timbered hills, the earth being a beautiful bright red, and in the afternoon came to, what seemed to us, the same stream we had left in the morning, thongh now much larger, crossed it, followed down some distance, crossed again and camped.

At both crossings the water was so deep we were compelled to undress, in order to ford it without getting our clothes wet; but the sensations produced by wading to the arms in such a current, breaking the ice, and climbing out on a snow-covered rock, and dressing in a frosty atmosphere, can only be known by experience. Any way, after the last crossing we danced around our blazing camp fire a long time before we ceased to shiver.

Last night there came a storm of sleet, after which the air became very cold and continued so all day. Crossing some mountain-like hills, about noon we came, as we supposed to the north fork of Feather river, near the forks.

Finding a place where the current ran deep and smooth, when our preparations were complete we hurriedly undressed and pushed through, finding it not only cold, but difficult and dangerous. However, we succeeded in getting all our "traps" safely over, but with the feeling that another such effort might be fatal to us both. And when afterward we returned to the same place we dared not attempt to cross.

It was our intention to follow up the north branch of this stream, but were compelled, on account of its canons and steep banks, to ascend the mountain which terminated between the forks, and thus keep in sight of the valley of the north branch. Here again we struck an Indian trail in the snow, and pursued it until reaching the head of a wide prairie-like slope, where we camped.

As we came up, and along the mountain, we saw a line of marsh (cattail) flags tied together, and suspended from tree to tree. Beginning at a point near the river, ascending the mountain, and after following along the ridge for several miles, it diverged from our course, and we left it. It seemed too frail to mark a boundary line; doubtless it was the work of Indians, and we were greatly puzzled to know what it meant.

Monday, Feb. 10. At the dawn we were on our way, and late in the afternoon reached a small valley surrounded with a belt of large pines. Across this the snow was marked with Indian paths, and smoke arising from the woods at the foot of a mountain spur, revealed the place of their village.

Having been warned that these Indians were hostile, we hesitated about leaving the shelter of the timber, and yet realized that we were dangerously near the town. Putting fresh caps on our rifles and pistols, and closely scanning the openings around, we took a main path, so as not to excite suspicion, even though seen from a distance, and hurrying across the open flat, were soon again under cover of the woods.

Here we found a trail leading to a small brook; following up this, and walking in the water, where our tracks could not be seen, went about a mile up the creek. Then taking off our boots, and tying our pantaloons close to our ankles, so that our footprints in the snow resembled moccasin tracks, we crossed a low, timbered ridge, over a mile northward to another small brook. Here we wrung the moisture from our socks, put on our boots and followed up the rill several miles, into a deep canon, where we camped.

As we were compelled to have a fire, not only to warm ourselves, for the night was very cold, but it was necessary to dry our boots and socks, which had become very wet while walking in the water and snow, we built it in the most secluded place we could find; under an overhanging rock, at the side of the canon, in front of which a large pine had fallen from the cliff above. Peeling some of the bark from the dead pine, and laying it on the snow, furnished quite a good floor, and a comfortable place to spread our blankets.

Here we cooked the last of our flour; moulding it into five small biscuits. These, with about a half pound of bacon, constituted our entire supply of provisions. Each taking a biscuit and a slice of raw bacon for supper we lay down to sleep.

The proximity of the Indian village excited our fears, lest we had been discovered, and might be attacked during the night; and therefore our sleep was hardly as sound as usual. Toward morning a noise aroused me, but not fully; however, Donnelly gave me a shake, and said in a hoarse whisper, "They're coming, they're coming," and in an instant were on our feet, with rifles ready in hand.

The fallen tree furnished a good breastwork, the fire had smouldered into darkness, and so we stood listening in breathless silence. At first it sounded like the hurrying tread of many feet, coming into the canon a short distance above us; then there was stillness, and again a renewal of the noise.

When day began to dawn we crept along under the shadow of the cliff, and discovered that our fright had been caused by a small avalanche which had slid into the canon, and was followed, at intervals, by masses of snow and rock. Possibly, if we had not been fearing Indians, it would not have disturbed us.

Ascending from our canon we took a northwesterly direction over deep snow and among pine, fir and cedar trees of immense size. The air was piercing cold, and notwithstanding our constant struggle in the snow, we found it necessary to kindle a fire occasionally and thaw out.

This was easily done. The action of the wind generally cleared away the snow, leaving quite a space around the base of each tree; often, where the snow was deep enough, to the depth of eight or ten feet. So when we found a dead pine, we simply climbed down and set the moss and pitch on fire, and when warmed up climbing out pursued our journey.

During the day we noticed several landmarks, which had been described to us by Mr. Ward, while at Downie's diggings; and about noon looked down into the valley of the north fork of Feather river. A wide plain, "The Big Meadows," stretched far to the northward, and near the upper extremity of these "Meadows," we were to find our company.

Our exertions were redoubled and before night we had crossed the immigrant road, and were near the upper boundary of the plain. This was indicated by a dark line of timber, behind which arose an array of glittering hights, which we supposed was the summit ridge of the Sierra Nevada.

Sometime after dark we turned into a grove, near the river, kindled a fire, broke off a quantity of fir branches, spread them on the snow for a bed, lay down and slept soundly until morning.

## CHAPTER VII.

*Pass Big Meadows.—Disappointed.—Find Another Indian Village.—Difficult to Pass —Special Danger.—Answered Prayers.—Sunrise on the Mountain.—Last of Our Provisions Eaten.—Find the Landscape as Described by Mr. Ward, but No White Men.—Again the Unexpected.—Begin the Retreat.—Almost Hopeless.—Pass the Last Indian Town —Hailed by Indians.—Take Them With Us —Let Them Return.*

Wednesday, February 12. This morning, chilled by the keen night air, it was sometime before we were warmed up. With our utmost exertions, it was late in the afternoon when we reached the timber at the upper extremity of the Big Meadows.

Ascending a high, bleak point, in hopes of seeing some signs of our company, Donnelly, who was first on the summit, exclaimed; "There they are; there they are," pointing to several columns of smoke curling above a dense forest about a mile distant.

For a while we were delighted with the prospect of a plentiful supper and comfort ble night's rest. But all this vanished when, through an opening, we caught a glimpse of several Indian wigwams. We had found another Indian village, and were anxious to avoid the place which at first we had hailed with so much delight. The town was in the diection we wanted to go, and it was difficult to make a detour on either side without great labor and loss of time, owing to the nature of the surrounding country.

Following the ridge we were on would enable us to pass the town at a considerable distance; but we almost certainly would be discovered on the snow; even the night would hardly conceal us. We therefore resolved to go down into the timber and pass the village in the night.

It was now about sunset, and finding a place of shelter and concealment, we waited until the woods became dark; then taking off our boots tied them to our packs. Fortunately we each had an extra pair of socks; these we put on, drawing the worn ones over them; tied down our pantaloons and made our way.

Thinking it best to follow their paths in the snow, so as to avoid being tracked, we were brought nearer the villege than otherwise we would have gone, and at one time supposed we had been discovered. We heard the tramp and saw the dimly outlined forms of several Indians coming towards us.

Stepping closely to the side of a large pine we stood shoulder to shoulder with our rifles leveled toward the group. Escape seemed impossible, death inevitable, and our only hope, to die suddenly, and thus escape torture. We commended our souls to God. Donnelly was a devout Roman Catholic, and, in low breath, prayed fervently, "Jesus, Mary and Joseph, have mercy on us."

We had actually surrendered all into the hands of God, and He had mercy on us. The Indians turned on another path and disappeared among the trees. It seemed like coming back to life. Hope revived and we hurried on. About midnight finding a dead pine from which we stripped a large piece of bark, laid it on the snow for a bed; we dared not kindle a fire, but, eating a biscuit and a slice of raw bacon; greatly exhausted with fatigue and hunger, but grateful to God for our lives, we lay down and slept.

Thursday, February 13. Chilled and benumbed, we were awake when the first traces of dawn appeared. Hastily folding our blankets, in hopes of finding some sign of our company, we ascended above the timber line, frequently pausing to scan the woods and vales below.

Sunrise found us on the summit, between two cone-like peaks. The immense scrolls of snow, which crested the mountain, flashed in the red sunlight, and presented a scene beautiful and grand beyond description.

Here we ate the last of our provisions; one small biscuit and a morsel of raw bacon. It seemed rather to sharpen our appetites, but we were excusable for not eating more.

Then, from various positions we anxiously scanned the landscape. There was the cliff; clustered pines in the river bend; the rocky point with its five dead pines; all of which Mr. Ward had described to us; but there were no recent signs of white men.

Under other circumstances the scene would have charmed us, but now a strange fear of not finding our company began to haunt us. Descending to the valley of the stream, we followed up until noon; and still no traces of white men; all was silent as the grave, save the murmur of the river, and the sigh of the pine tops

Again we ascended a high ridge, and took another long, anxious look. The hills seemed solemn and stern, the dark lines of timber appeared cheerless. Far over the snowy ridges, we could see the towering summit of Mount Shasta, rising like a marble pyramid in the sky. Painful as was the thought, we were compelled to give up all hope of finding our company.

Whether they had been butchered by the Indians, or had crossed the mountains to Klamath river, we could only guess; but we were satisfied that for months they had not been where Mr. Ward left them.

Since approaching the immigrant road we found places where camps had been established and trees chopped; and so at the point where we expected to find our company; but there was no evidence that any white man had visited these places since the snow fell. Not having a list of their names, we were unable to inquire for them personally in the Klamath, Pitt river, or other mines where they might have gone. A strange mystery enshrouded their fate, and what became of them we never learned, but strongly suspected that the whole party of eighteen were surprised and killed by Indians.

But, what were we to do? The unexpected had happened. Entirely destitute of provisions, and already weakened with hunger; when we thought of the long distance to the nearest mining camp of which we knew, there seemed scarcely a hope that we could make the return trip.

However, as every hour of delay rendered escape less certain, we turned back, aud followed down the river with all possible speed. It was a journey for life, and we did our best.

Stripping some bark from the sugar pine we chewed it while passing along, and sometime between midnight and morning we passed the upper Indian village, without paying much attention to the pathes, and at the edge of the big meadows crept into a jungle of drooping firs to obtain a few hours rest and sleep.

In all our journey there has been an unaccountable absence of game. When we began our journey wolves were generally within hearing at night, but we have not seen a deer, pheasant, or even a rabbit, though for several days we have been constantly on the watch for them. Possibly the deep snows had driven them from this region, or it may be that the Indians had taken all the game within their range.

We were not prepared, with hook and line, to try the streams for fish, but wherever we examined the rivers, had failed to find any, and, consequently, were not able to add anything to our original supply of provisions. We would gladly have slaughtered even a wolf for our supper, but in our extremity, they too, kept strangely out of our way.

Friday, Feb. 14. Although suffering extremely from hunger, we walked rapidly all day, and about sunset entered the timber at the lower extremity of the Big Meadows. It was our intention and hope to pass the Indian village during the night, but we were too tired to continue the journey.

Fully conscious of the danger of remaining near the village, or trying to pass it in daylight, yet, in utter exhaustion, we crept into a clump of firs, and slept several hours.

Before day the journey was resumed. Our sleep rested us somewhat, but the chill had so stiffened our joints that for awhile we made but little progress. However, in the morning twilight we safely passed the village, and in about a mile below, struck the trail by which we came up the river.

Beginning to feel relieved of danger from Indians, and congratulating ourselves that escape now depended only upon our physical endurance, when, lo, we were hailed; and there, only a few rods in front, where our path swayed to the left, at the head of a ravine, up which, evidently, they had just come, stood two tall Indians, making signs for us to approach them.

"Don't let them think we are afraid," said Donnelly, and we promptly started toward them; instantly agreeing that we must make them travel with us that day, and not permit them to report us at their village, if we could possibly prevent it.

They were armed with bows and arrows, with small hunting knives in their belts, and as we came near, made signs, pointing to us, and then to the path, in the direction toward their town.

Donnelly with one hand upon his pistol, with the other pointed first toward them and then to the path in the opposite direction. As we expected, they were inclined to resist. One succeeded in placing an arrow in his bow, but before he could raise it, Donnelly surprised him by presenting a pistol near his face. He understood what it meant and dropped the arrow.

At the same instant, before the other had placed the arrow in his bow, but held it in his right hand, anxious to avoid the report of fire arms, I drew my hunting knife and grasped him by the shoulder; but Donnelly, fearing that he might seize me, leveled the pistol at his head.

With a quick movement, I drew a pistol, stepped back and leveled it. Dropping the arrow, they both stood for a moment as if undecided and angry but seeing that we had them in our power, they turned, and talking to each other took the path, and we, putting up our pistols, followed a short distance behind.

Presently they stopped, and evidently, were about to raise a yell, but our rifles were quickly leveled upon them, and again they turned and pursued the path before us.

They kept watch of us every moment, and several times slackened their pace, and acted as though they intended to turn upon us; when at the click of our locks, as we prepared to shoot, they would increase their pace, but they compelled our most diligent and constant attention.

In our weakened condition, it was a terrible strain upon our nerves, for there was in our minds the inexpressible dread that we might have to shoot those Indians in order to save our own lives. The awful specter of death which haunted us, seemed to take away our sense of hunger and weariness, and no doubt that under the excitement we traveled further than otherwise we would have done.

Near sunset we gave them to understand that they might return, which they did with an apparent good will.

We were sorry to compel them to travel all day without anything to eat, but in that respect we all fared alike. We would gladly have given them a good dinner, and valuable presents to remember us by, but were pleased that our relations had been no more unfriendly.

We would rejoice to have gained their good will, but under the circumstances it seemed impossible, and it is doubtful whether the next white men which they met, were treated with the same consideration bestowed upon us.

However, fearing lest they might lurk on our trail and attack us during the night, from the shelter of a thicket we watched their retreating forms until they disappeared several miles distant.

After this we traveled about five miles, and, obtaining a supply of pine bark to chew, crept into a jungle in hopes of finding rest and sleep.

## CHAPTER VIII.

*Strange Illusions.—Dare not Venture into the River.—Follow down the Bank.—Mountaineer Lawson's Mining Camp.—Obtain Provisions.—Almost a Quarrel.—Nausea.—Prospecting.—Snow Storm.—Start for Downie's Diggings.—Disagree.—A Terrible Moment.—Separate.—Together Again.—Camp at the American Ranch.—A Pocket full of Money, and Didn't Know it.—Onion Valley.—Cooking under Difficulties.—Luxury of Sleeping with Boots off.—Under a Tent-full of Snow.—Getting our Boots on.—Hotel at Onion Valley.—Comfortless Crowd.*

Sunday, Feb. 16. With the first traces of dawn we were again on our way. Did not suffer so very much from hunger, but, as we warmed up with exercise, felt feverish, and, for awhile, traveled with comparative ease.

I noticed that Donnelly's eyes were swollen and blood shot, and at times we staggered from the path.

Passing near the brink of a steep ravine, Donnelly remarked, "Very likely there is gold down there," and pausing, we seemed to hear the sound of human voices. Listening, the tones were very distinct, but we could not distinguish the words. Then came the sound of digging with a pick, the scraping of a shovel on the rock, and the peculiar noise made by shoveling gravel into a tin pan.

Confident that a company of prospectors were at work in the ravine, we at once descended; not doubting but they could afford us something to eat.

However, we failed to find the slightest trace of any human being. The snow lay undisturbed even by the foot of a rabbit, and the voices and sounds were purely imaginary.

Listening, we heard them again, just above us, in a bend of the ravine, but when we reached the place, there were no signs.

For some time we were inclined to follow the illusion; our dismay was inexpressible; we did so want to make it true, but at last realized that we were only wasting time, climbed out of our ravine, and, fortunately, found our path.

Strangely enough, both of us seemed to hear whatever we listened for. Sometimes it was human voices, but never an atriculate word, and sometimes it was the clang of mining operations. Several times we were tempted to turn from our path, and it was difficult to realize that the sounds were only illusions of the mind.

About noon descending to the river where we crossed when coming up, but, looking at the distance, and the swift, angry current, realized that it would be impossible, in our exhausted condition, to cross in safety. We therefore turned down the river, hoping, at least, to find a safer crossing.

However, in several miles, we were surprised and delighted to find mountaineer Lawson's mining camp. There were several white men and Pah Ute Indians working a placer mine on the bank of the river.

Coming down the river, we saw that what we had taken for the valley of the north fork of the north fork of Feather river, was only a deep gulch, and the main stream, which we had visited above the Big Meadows, lay over the mountains toward the northwest.

From Mr. Lawson we obtained some flour and bacon, and in a few minutes I had a cake baking in one of their skillets. While waiting for the cake, Lawson, who was familiar with all that region, inquired of Donnelly about our journey, which was candidly described. Whereupon Lawson bluffly replied, "Don't tell me any such stuff as that; I know that country; it's not far from a hundred and sixty miles; you fellows never made that trip without eating."

In a moment we were very angry, and Donnelly laid his hand upon his pistol; but Lawson repeated, "Don't tell me any such stuff." He evidently did not believe what Donnelly had said, and we, having our veracity questioned, and our honor contemned, were at once on our dignity, and in no condition to reason or explain, and as we had already paid for the flour and bacon, indignantly refused to have any further conversation with him.

After eating a little, we both experienced nausea, and for awhile were unable to travel; gradually, however, the feeling passed away, and gathering up the fragments of our dinner, we hurried away, and camped at some distance down the river.

By occasionally eating a little our stomachs regained their natural tone, but it was several days before our mental vision became clear.

The next day we did some prospecting along the river, and finally, crossing on a large pine which had been felled, camped with a company of prospectors.

Tuesday, Feb. 18. Our long spell of fair weather terminated last night in a snow storm, and this morning, in the open camp, found us literally snowed under, and the snow still falling. We had scarcely more provisions than were needed for breakfast, but on our way back to Downie's diggings we expected to replenish our stock at the American Ranch, a place somewhere on the way to the middle fork of Feather river.

Ascending the mountain among large trees, everything enveloped with the same white mantle, with neither sun nor outlined hill in sight, we wallowed through the snow most of the forenoon, and finally disagreed as to our course. Whether going east or west, neither could be absolutely certain, but each felt sure he was right. Donnelly had our pocket compass, and I wanted him to examine it, but, he was so sure of being right, he refused; and finally I told him I would not go in that direction any further. Angry words followed, and suddenly he aimed his rifle at me. In an instant I leveled a pistol on him. Thus we stood for a moment. The Good Providence prevented our shooting.

Donnelly was first to understand the situation, and threw his rifle into the snow, where it sunk out of sight. As he picked it out of the snow, I knew it was too wet to fire, so putting up my pistol, turned away, and there we parted.

I felt very badly. Hot tears trickled down my cheeks. We had crossed the plains together, had always been fast friends, and no shadow had ever before darkened our brotherly love. Strangely bewildered, I went on in the direction which I supposed led to the American Ranch.

In about an hour Donnelly again appeared, converging toward me, and when we met, he simply said, "I believe this *is* the right direction;" and we never mentioned to each other our terrible episode. But words could not express my joy in the consciousness that we were still friends.

I learned through our mutual friend, Robert McCord, that after we separated, Donnelly looked at the compass, and of course, it was the merest accident that I happened to be right; but had we slain each other there in the woods, it has seemed to me, that, because our minds were so unsettled, but little moral responsibility for the deed could have attached to us.

So too in our intercourse with Lawson; had we been able to have explained our situation calmly to him, he might have been of great service to us; but he, no doubt, thought we were trying to impose on him; and we, conscious of telling the truth, felt too keenly his expressed disbelief of our story, to attempt any explanation.

Late in the afternoon the snow slackened some, and about dark we reached the American Ranch; a large log house at the edge of an open valley. Here we obtained some provisions, and fixed our camp under the branches of a large, low pine, on the bank of a clear brook.

Wednesday, Feb. 19. Last night when our hot coffee, bacon and bread were ready, a young man came up from the American Ranch with the information that he was "plumb strapped," which was miner's parlance for saying that he was entirely destitute of both money or provisions. Our hunger had been too recent to permit us to be unmindful of his need, so we gave him a cordial invitation to take supper and breakfast with us, which he did, and finally concluded to accompany us to Downie's Diggings on Poorman's creek, and join us in a mining enterprise.

In the morning, while waiting for us to prepare a lunch for dinner, he lay down on his blanket by the fire, and in an unguarded moment turned on his side, when, out of his pocket rolled a handful of gold and silver coins. Hastily gathering them up, he remarked, "*I didn't know I had it.*"

The idea of a person carrying such a weight in his pantaloons pocket and not knowing it, brought to our faces an incredulous smile, whereupon he seemed embarrassed, and at last, without even saying good bye, left in the direction of the American Ranch.

No effort was made to persuade him to remain with us, and carry out our mining project. Even without his falsehood, his pocket full of coin indicated that he was a professional gambler. Gold dust is the currency of the mines, but a gambler always keeps a supply of coin as an attractive display, and should he lose in his game, always pays in "dust."

We crossed the middle fork of Feather river at the mouth of Nelson's creek, and thence over a high mountain, in a heavy snow storm to Onion valley, arriving at the only buildings, two overcrowded cabins, about dark. However, a prospector kindly permitted us the use of a tent.

Here we mixed and kneaded a cake for our supper, and undertook to bake it at a fire built outside against a large pine log. But the constant high wind showered the snow upon it, and prevented the formation of the usual delicate crust, and after holding it awhile in the flame it was pronounced "done," and although stained with smoke, and tasting of pitch, it satisfied our hunger, and that was all the very best could have done.

Thankful for shelter, we enjoyed the unusual luxury of sleeping with our boots off. Something we had not done for weeks, owing to the cold, and our exposed condition. It was a great relief to our weary feet, and we slept well.

There was a heavy gale during the night, and morning brought no abatement of the storm. The tent door was buttoned in the center, but the wind had burst off the lower button and had made a rent in the opposite upper end, where it found exit, and had piled the snow from the bottom of the door to the peak of the tent. We lay under the drift, our heads only projecting from it near the door, and sheltered by our caps. Our boots, somewhere under the snow bank, when found were so frozen, being wet when taken off, that they had to be thawed before we could put them on.

Fortunately, within about ten rods, several men were trying to save their horses and mules, in the shelter of the woods, by the aid of a great fire. Thither we waded the snow barefoot, put on our socks, thawed our boots, and finally succeeded in getting into them.

Next going to one of the large cabins, we inquired of a man behind a bar, at the side of the room, whether we could obtain breakfast, "Yes, sir, breakfast, or any other meal you want, just as soon as you can get a place at the table."

A narrow puncheon table, probably twelve feet long, occupied the center of the room; on either side a narrow puncheon seat, the length of the table, and, like it, resting on posts driven firmly into the ground, which constituted the floor. On each side of the room were shelves resting on large pins, projecting from auger holes in the logs, and furnishing a receptacle for provisions, liquors, etc.

Against the logs at one end of the room, was a thick wall about eight feet long and four in hight; in front of this the kitchen fire was built, the smoke finding exit through a large hole in the roof. Two men, engaged in cooking, were scarcely able to supply the eaters which thronged the table, while the hungry crowd around waited impatiently their turn.

It was nearly noon, when after weighing out in gold dust, three dollars apiece, Donnelly and I, crowded together on the puncheon table seat, were furnished with a cup of coffee, a slice of fried bacon, and a piece of bread broken from a cake, which had been taken hot from before the fire; but it was all delicious beyond expression. Toward night we obtained another meal.

There were probably two hundred trying to shelter ourselves at this unfinished hotel, and it was difficult to prevent being in each others way. The storm continued unabated, and the snow was full higher than the roof of the house, so that in going out we ascended a steep hill-like bank. Considerable snow was tracked in, and, dissolving on the ground, our standing room became a pool of mud.

When night came on, and we all crowded in, it was simply impossible for all to lie down. Donnelly and I wrapped our blankets about us, and sitting on a piece of fire wood which lay in the mud, slept as best we could, but, in common with others, were greatly annoyed by several drunken men, who so disturbed the company, as at times, to threaten a first class tragedy.

## CHAPTER IX.

*Drink or Die.—Benefit of the Pledge.—One Meal a Day, but Plenty of Whiskey.—Delirium Tremens.—Man Lost.—Found Severely Frozen.—Dies.—Struggle with a Maniac.—Dying Men.—Drunken Revel.—Leave Onion Valley.—Mule Trains on Snow 40 Feet Deep.—Return to Poorman's Creek near the Yuba.—Our Former Partners Discouraged.—Prospecting.—Klamath.—Nevada City Burned.*

Friday, Feb. 21. The storm continued unabated all day, and we kept closely indoor.

This afternoon an intoxicated man invited me to drink with him, and I declined as politely and pleasantly as possible. But he was not satisfied, and began to insist; still refusing, I tried to move away, when he exclaimed, "Hold on, .you think I'm drunk, and are ashamed to drink with me, but I'll make you do it;" and seizing a bottle, poured a quantity of liquor in two glasses that stood on the bar, pushed one toward me, and, at the same time producing a revolver, remarked, "There now, take that glass of liquor, or the contents of this pistol." He evidently meant, Drink or die.

"Wait a moment," said I, "my partner can explain this."

Donnelly was called. "Here is Mr. Donnelly, my partner; we crossed the plains together, he knows whether I ever drink with any one; now, Mr. Donnelly, did you ever see me drink with any one?"

"No sir, I never did."

"Did you ever hear me give a reason for not drinking?"

"I've heard you say you were pledged against it."

Now turning to the man, who had put up his pistol, I said, "A gentleman like you would not ask any one to break his word."

"Of course not, and here's my hand on it."

So we shook hands, and the affair was

settled; but I was more than glad of being pledged against drinking.

Saturday, Feb. 22. The day was very cold with a fierce gale, occasionally tearing the clouds and letting through a ray of sunshine. This morning our host informed us that he had consulted the proprietor of the other hotel, and it was ascertained that the supply of provisions was nearly exhausted, and we would therefore be limited, each, to one meal a day.

Before night I could not refrain the wish that the liquor might be reduced to even less than one drink a day; but however scant our ration of bread and bacon, there was plenty of whiskey, and the place became a veritable pandemonium.

In the afternoon a large German was attacked with delirium tremens, and became so violent that the men bound him, hand and foot, and laid him, a shrieking maniac, in one corner of the room; but I don't know as he made more noise than some who had not yet reached that stage in the drama, and were only drunk.

After dark, a man came in and told us that his partner was somewhere out in the woods. Being intoxicated they had started for Downie's diggings on Poorman's creek, but lost their way, and finally one succeeded in getting back to the house.

Quite a number started out for the missing man. We divided in squads, three or four together. It was a fearful night; so very cold, and a fierce gale scattered the branches, making it dangerous to be in the woods, on account of the falling limbs.

He was found by our squad, in a depression in the snow, leaning against a large pine, unable to speak, his face frozen, and icicles hanging from his bare hands. Rolling him in a blanket, and carrying him back, as we neared the house we were joined by Dr. Y., formerly a surgeon in the United States army, who, after feeling of his face and hands, told us to take him down to the spring, where he could be laid in a bed of earth until the frost was withdrawn.

Unfortunately, the news that he had been found preceded us, and when we stopped at the hotel for a pick and spade, the rabble came out and insisted on taking him in to the fire. We urged the necessiity of following the doctor's advice, and for awhile there was a fair prospect of a fight, but when they drew their pistols we were forced to give him up, and he was taken in, and laid upon a shelf near the fire. The drunken mob had its own way, but the poor man died in the morning, after a night of terrible agony.

The German, afflicted with delirium, falling into a comatose condition, had been unbound, and was lying quietly on some clap boards placed on the mud floor It was near midnight, and Donnelly and I, hoping to obtain a few hours rest, brought in some boards, which we found piled against the house outside, placed them on the mud in a corner of the room for a bed, wrapped our blankets about us, lay down, and were soon asleep.

About that time the German aroused with a scream, and when I awaked, he was kneeling beside me in the mud, and reaching across to Donnelly, who lay next the logs, had him firmly by the throat. I tried to open his hands, but his grasp was like an iron vise.

Seeing that Donnelly was choking I called for help, and some of the revelers yelled back, —— him, why don't you shoot him?"

In a moment his hands relaxed, and rising with a bound, he ran screaming against the table, almost pushing it off the posts, but, soon overpowered, he was again bound and laid aside, and his screams gradually died away in sob-like groans.

Donnelly had been choked almost into insensibility, but as I raised his head and shoulders, his breath returned. However, for several days he felt the effects of the maniac's fingers upon his throat.

Words utterly fail to picture the scene in that cabin. Above the drunken revel at times could be heard the pleadings for help of those two men. And the doctor, who, had he kept sober, might have rendered some help, became wild with drink, and after singing vulgar songs for awhile, finally quieted down into a drunken sleep.

About sunrise the man who had been frozen breathed his last; the one with delirium lingered through the day, and as I afterward learned, died during the following night.

Donnelly and I determined to leave the valley; three days and nights in such uncomfortable quarters were enough for our patience; and besides, the immense depth of snow rendered mining at Downie's diggings impossible for some time to come. Those who had claims already proved to be rich, could wait in hope, but to us it seemed better to spend the time in active prospecting.

Therefore, as soon as we could obtain our daily meal, which was early in the afternoon, we started for Grass valley, said to be nine miles distant.

From Onion valley we ascended a ridge where the snow had almost completely covered the forest, and yet, on the top of this great depth of snow, the enterprising owners of pack trains had beaten a path so that horses and mules with their loads could travel in safety.

These trains continued to travel until the beginning of the recent storms, and were only discontinued when the severe freezing, high winds, with added snow, made the path dangerous. But I learned later, that although the storm continued for a week after we left, as soon as it abated, the wind had ceased, and milder weather came, the path was beaten out, and pack trains again traveled over the snow.

It would be difficult to guess the average depth of snow, but Mr. Christopher R. Stark, whose home has since been near Granville, Ohio, and who that winter and the following summer was engaged in packing provisions, with a mule train, over that trail, informed me that a limb, over which they had traveled for some time, when the snow began to settle became an obstruction, and was cut off, and when the snow was gone, the branch from which that limb was cut was found to be forty feet above the ground, his train passing over snow of that depth.

At Grass valley we found an overcrowded hotel, but succeeded in obtaining supper, and passed the night in comparative comfort.

Monday, Feb. 24. This forenoon there fell a light snow. It seemed to be one of the outlying curtains of the storm which still enveloped the great hights from which we had come. We made a fair day's journey, and about night stopped at the Buck Eye ranch. We were now out of the region of snow, flowers were in bloom, the air balmy as an evening in spring, and yet only a day and a half from where winter reigns in all its rigour. A most agreeable contrast for us.

Having a desire to know how the mines had developed in the valley of the South Yuba, we returned as fast as possible, and found our former partners, Hunt, McCord and Farley, at our cabin on Poorman's creek.

Their mine had scarcely paid expenses, but they had worked on, hoping it would improve; it, however, had paid less and less, and now that it was nearly worked out, they were ready to quit.

This was a common experience, all through the mines; hundreds of men worked hard, early and late, encouraged by the hope, never realized, of finding a rich deposit.

Wednesday, March 5th, 1851. This morning we started in search of new mines. Donnelly, Hunt and Farley explored the South Yuba, while McCord and I examined the gulches and creeks between the Middle and South Yubas. The weather was pleasant, and after three days diligent but fruitless search, we crossed the South Ynba, and followed up Deer creek to Nevada City.

Here we learned that our friends William E. Shimmans and Henry Callanan had gone to the Klamath mines, which had been reported very rich. John Callanan was arranging their business affairs, intending to go when he heard from his brother Henry, provided the story of the richness of the mines proved true.

Of course this was interesting to McCord and I, and we felt inclined to keep within hail of Mr. Callanan until the question of the Klamath mines was decided.

The mines around Nevada had produced an immense quantity of gold, and the business of the town had greatly extended, but just then the people were in dread of a band of roughs who had threatened to burn the city.

The saloons and gambling houses had developed and sheltered a vicious class, which became so numerous and desperate; emboldened by the lack of organized government, they greatly interfered with legitimate business, and men known to have gold or other valuables were in constant danger of being murdered and robbed.

The better element combined in self-defense, and demanded that certain known and designated criminals must leave the place within a specified time, or be put to death. They left, but with the counter threat, to return and burn the city. And so for some time the people had been in terror lest the threat might be carried out.

Returning to our cabin on Poorman's creek, we found the rest of our company. Our prospecting had been unsuccessful, and we made up our mind to go to Nevada, and perhaps, ultimately, to the Klamath mines.

Tuesday, March 11. This morning taking our effects on our shoulders, we started for Nevada. The day was very warm, and our burdens heavy, and about dark we reached the ridge which overlooks the city from the north. Here we camped, near Sugar Loaf hill.

About midnight, one of our company discovered that the city was on fire. Beginning in the valley of Deer creek, in the south east part of the city, the fire soon communicated to a large store. A high wind from the south east swept the burning coals over the canvass covered buildings, and in a short time the whole city was one mass of flame.

The houses, built of wood and canvass, were soon gone, leaving a smoking mass of ruined merchandise, and a large number of homeless people. The loss was estimated at about four hundred thousand dollars.

The fire had been kindled in a ball alley, evidently by the banished thugs, and in fulfillment of their threat. While contemplating the terrible deed, many, maddened by their losses, expressed regret that they had not been put to death while they had them within reach, there by not only saving the city, but preventing the repetition of similar crimes against others,

The destruction of such a quantity of provisions produced something of a famine, until supplies could be brought from Sacramento; but, fortunately the roads were quite good, and in a few days there was a city of tents, and abundant stores, and soon the city began to rise again, with better arrangement of streets, and more substantial buildings.

## CHAPTER X.

*Sluicewashing at Nevada.—Rain and Snow.—Favorable Report from the Klamath.—Start for Klamath Mines.—Journey to Sacramento.—Efforts to Reach the Post Office.—Another Vain Effort.—Sutter's Fort.—Delegated to Visit the Klamath Mines.—Sacramento Post Office.—Letters from Home.—My Company Return to the American River.—I Sail for San Francisco.—Delay.—Vessels Abandoned by their Crews.—Deception, Crime and Suffering.*

While waiting news from the Klamath we engaged in sluice washing, hiring water from the Deer Creek Water Company, which had succeeded in bringing it in a small canal from near the head of the creek onto Coyote ridge.

Our mining arrangements were very simple. A ditch about eighteen inches wide and twelve deep, and ten rods long, was made in the ground, terminating in a longtom and riffle-box. Into this ditch was turned a stream of water, one inch deep and five wide, under a pressure of one inch head; and for this five inch stream of water we paid five dollars a day.

Into this shoveling the gravel, raised from the deep mines around, which, though containing considerable gold, was not sufficiently rich to pay for hauling to the creek, but lay in vast piles of reffuse. This while running down the sluice and longtom was thoroughly washed, the gravel shoveled away, the sand running through, and the gold being heavier, remaining in the little eddies in the riffle box.

After working thus all day, the sluice was carefully washed down, and all the gold collected in the riffle-box and then "panned out." That is, placed in a large tin or iron pan, and the remaining sand carefully washed away. Thus, after paying five dollars for water, we obtained from five to seven dollars each per day.

Wednesday, March 19. Rain fell during the entire day, and the next morning, snow, to the depth of eight inches covered the ground; and for two days, snow, sometimes mingled with rain, continued to fall, but there was little or no frost.

Saturday, March 22. This morning the sky was cloudless. The bright sun and balmy air soon dissolved the snow, which settled quietly into the ground.

We finished our sluice washing, and receiving a favorable report from the Klamath mines, prepared to go there, by way of San Francisco.

Monday, March 24. This morning Thomas Hunt, Drury Farley, Robert McCord, John Donnelly and myself started on foot for Sacramento city. The country through which we traveled, gently undulating, with the rich foliage of trees, carpeting of grass and flowers, possessed great natural beauty; and when from a spur of the foot-hills we obtained a view of the Sacramento valley, spreading out in a vast green meadow, it reminded us of our journey last summer along the Platte.

We spent the first night at Union valley, and the second day at dark ferried the American river, stopping at a hotel about a mile from Sacramento city.

The next morning we entered the city, where Hunt and Farley, who are half brothers, hearing of their brother, John Farley, requested us to remain while they made him a visit. Through the kindness of a relative of Mr. Hunt, Mr. Adolphus Hanna, of the firm of Hanna, Jennings and White, largely proprietors of Sacramento city, we obtained a room where we could keep our effects and sleep; and thus, while living cheaply and pleasantly, have an opportunity of seeing the city and surroundings.

Our first visit was to the Post Office. I am quite sure that at this time there was not a post office in the mines. Letters for miners were addressed to Sacramento, and, of course, the mail arriving here was immense, and when we reached the office the crowd was too great for us to approach the delivery during the day.

The next morning we were there an hour before the time of opening, but the crowd seemed just as great as ever, so we retired again, and spent part of the day in visiting Sutter's Fort.

It stands about a mile from the Sacramento river, and was built many years ago by John A. Sutter, a Swiss by birth, and formerly a captain in the French army. The thick outer walls, with bastioned corners, on which are places for canon, were built of adobe, (sun dried brick,) inclosing a space of about twenty rods square.

This outer wall, fifteen feet high, is separated twenty-five feet from an inner wall ten feet high. This space was roofed over, making a tereplein protected by a parapet. Underneath were rooms for barracks, shops, stables, etc.

Quite a pretentious frame building stood within the inclosure, said to be the former residence of Captain Sutter. The only access was by two massive gates, one on the south, the other on the east; but they needed repairs, and like the dilapidated and deserted barracks, and dis mounted canon, disclosed the fact that military occupants and discipline had departed.

Saturday, March 29. Mr. Hunt returned this morning; said that he and Farley would not go with us at this time; but they, with McCord and Donnelly, wanted I should go, examine the mines, and report to them by letter at Mormon Island.

The fact that my friends Shimmans and Callanan had been there long enough to test the mines, induced me to accept the mission. However, we re solved to make one more effort to get our mail.

The condition of this post office is altogether unique. It opens at eight in the morning, and closes at eight in the afternoon. There is a delivery window for nearly every letter of the alphabet, and at each there is a row of people, often reaching more than around the block. When so many come in person for their mail it is simply overwhelming, and when it comes time to close the office, the lines break up, each to take his chances another day. But as hope deferred makes the heart sick, so, many who came a great distance and waited long, are compelled to turn away still enduring their anxious suspense.

Recently people have adopted the plan of having their mail addressed, "By express to Nevada, Coloma," or wherever they may be. Thus the postmaster at Sacramento can send the mail, by responsible express agents, to the various mining towns, and greatly relieve the office.

Monday, March 31. This morning about one o'clock we arrived at the post office, and found a large number in waiting. The line facing the S delivery window, already extended half way around the block.

Taking my place in the line, I waited until the office opened, and as the line in front melted away, moved forward. Of course each one of our little company sought the delivery according to name. This put us into different lines, and as we approached the window men came and tried to buy a place in the line, offering twenty-five and fifty dollars, and I was told that even a hundred dollars had been paid for a place near the delivery.

The one who sold his place, stepped from the line and went to the extreme rear, or else waited until the office closed and night had shortened the line, and again found a place. Many, who were near the delivery when the office closed, remained, holding their place until it opened in the morning.

At last I reached the delivery, and the busy clerk, after looking over a vast pile of mail matter, handed out what belonged to me. Gladly I got out of the way, and hurrying to our room, scanned the familiar writing, and with a strange tremor read the first letters I had received from home and friends since leaving them more than a year before.

My companions also received considerable mail, and we spent a portion of the day in answering letters; and in the afternoon they started for Mormon Island, on the south fork of the American river, about twenty-five miles from Sacramento.

Tuesday, April 1st, 1851. At two in the afternoon I left Sacramento on the steamer Wilson G. Hunt, arriving at San Francisco early the next morning.

Calling at the general shipping office, I was informed that one vessel would sail that day for Portland, Oregon, but as I was the only one who had applied for passage to Trinidad bay, the landing place of those going to the Klamath, they were not willing it should stop there.

Having leisure I looked around the city and bay, and was greatly surprised at the number of vessels lying in the harbor, but learned that many were there because they had been abandoned by their crew, the gold mines having tempted the sailors to desert, and the officers unable to obtain others, were compelled to remain.

I was told that on the 22nd of February ships of every nation on the globe were in the harbor, and displayed their flags in connection with the United States flag in honor of Washington's birthday.

Friday, April 4. This afternoon the Sir Charles Napier, a large English merchant vessel from Panama, arrived at Long Wharf, bringing a large number of passengers, all in a very debilitated condition, some of whom stopped at my hotel, the Atlantic, and related a terrible story of deception, suffering and crime.

A company in New York city advertised to take passengers from New York to San Francisco at considerably less than the rates charged by the regular steamers. Passengers were to furnish their own transportation across the isthmus from Chagres to Panama, where ships were promised to be in readiness to carry them to San Francisco. Hundreds accepted the offer and were landed at Chagres; made their way across the isthmus, and waited at Panama for the ships that never came.

At last, realizing the fact that they had been deceived, that no arrangement had been made for their conveyance further, some obtained passage on the regular steamers to San Francisco: others after long delay went on merchant vessels; and many died.

The New York company after carrying on its deceptive work until fearing detection and arrest, canceled the contracts for the ships they chartered, disbanded and disappeared.

About this time the Sir Charles Napier, an English merchantman, came to Panama, and taking on board all the passengers that could be accommodated, sailed for San Francisco. While on the way they were becalmed in the tropic for some sixty days.

Water and provisions failed, ship fever set in, many died, and, after a voyage of over four months the ship reached San Francisco. Among those who were buried at sea, were two whom I had known in Delaware county, New York; Walter Rutherford, a neighbor, and Garret McFarland, a schoolmate.

Through the incidental conversation of strangers, the well remembered names came to my ears. "Died and were buried at sea," was all they could tell; but how well I knew them both. Rutherford, almost a giant in stature and strength, once while whetting a scythe in his field was struck by lightning, suffering for months a mental and physical collapse, but rallying, appeared strong and healthful as ever.

When a small boy, the first day of my school life, Garret and I were placed in the same class, and for several years, summer and winter, our lives lay parallel; studies and interests seemed identical. Never but once, was there the least unpleasantness. Then, in a scuffle, I made his nose bleed; but perhaps it hurt me as much as it did him; and I went with him to the spring to wash away the blood.

And when school was called, possibly some one told the teacher that we had been quarreling, for he noticed blood on Garret's cravat, and inquired about it. "Yes," said he, "I got my nose hurt, and it bled a little."

I was not blameless in the matter, but Garret was just as anxious to shield me from blame as though I had been. Brave, generous, noble-hearted boy.

## CHAPTER XI.

*People Leaving the Klamath Mines.—Discouraging Reports.—Depart for Sacramento.—Caught in a Gale.—Annoying Experience.—Rejoin my Company at Mormon Island.—Mining with the Rocker.—Organize a Company for Draining the River Bed.—Rheumatic Fever.—Mass Convention Makes Mining Rules.—Peculiarities of Climate.—Death of the Stranger.—"Help here Quick!"*

Saturday, April 5. This morning two steamers arrived from the north, both stopping at Trinidad bay, brought a large number of passengers from the Klamath mines. All with whom I conversed gave a very unfavorable account of the mines in that district.

This was discouraging; and as the regular steamer would not return to Trinidad for fifteen days, I resolved to report at once, and in person, to my company at Mormon Island, and at four in the afternoon took passage on the steamer West Point for Sacramento. A little after dark, while crossing the bay, our boat came near being lost in a gale, which swept everything from the decks, and put out the fires.

After the force of the storm had passed, a sail was raised, and toward morning arriving at Benicia, the boat was repaired, and the fires rekindled, but it was late Sunday night when we reached Sacramento.



Being without sleep the preceding night, I went to a hotel, and slept soundly until morning; but waking, felt a strange creeping sensation on all parts of my body. An examination revealed the disgusting fact that I was infested with insects of the species *pediculus vestimenti*, or greybacks. I had heard of such things, but had never seen them before, and the experience was decidedly annoying.

However, after breakfast I went to a clothing store and bought a complete suit, including underwear, folded them in my satchel, and then started on foot for Mormon Island. In a few miles reaching the American river, where, sheltered by a grove, I undressed, taking my clothes, stockings and all, rolling them in a bundle, sunk them in about two feet of water, placing a large stone on top, and for aught I know they are there yet.

Then, after a thorough washing with sand and soap, I donned my new outfit and felt no further annoyance. An easy walk of twenty-five miles, and, late in the afternoon, I found my company camped a little above Mormon Island, on the south fork of the American river. They reported it a very rich mining district, and we were all pleased to give up the journey to the Klamath.

We learned afterward, that while a few very rich discoveries were made on the Klamath river, they were not extensive. But the transportation companies, by extensive advertising induced an immense rush to that region.

It was a rich harvest for the shippers, but hard on the miners, who, after spending the time and labor of exploration, and paying their passage both ways, returned to their former diggings, under all the disadvantages and losses which follow a break up in business.

We now adopted the use of the "rocker" for gold washing. A machine made in various styles, but in general outline like a cradle in which babies are rocked. The part corresponding to the head, has a box about five inches deep, with a sheet iron bottom perforated with half inch holes. Into this a bucket of sand and gravel is thrown, and, while with a dipper, water is poured over it, the cradle is violently rocked, and when washed clean, the gravel is thrown off, and another bucket supplied.

On the inside across the bottom, slats were nailed, forming ripples where the gold may settle, and permit the sand and light gravel to run over. This necessitated placing the rocker at such an angle as to make the water ripple just right, so as to let the gold settle and the sand run off.

It is a slow method, but enables a person to use water from a pond or river, where it cannot be raised so as to run into a sluice or longtom. With the rocker each one generally worked alone, and we were enabled to wash little bars and banks, many of which were exceedingly rich. In this way we did a profitable business; sometimes collecting twenty-five or thirty dollars worth of gold in a day; but a half ounce, eight dollars, was considered a fair day's work.

Monday, April 21. To-day R. McCord and I, with twelve others, organized a company for the purpose of turning about forty rods of the south fork of the American river from its channel, in order to obtain the gold from its bed.

In connection with other companies, we selected our claim about half a mile below Mormon Island, arranging to take the water of the river from a large ditch belonging to the next company above, and bringing it over the river bed, into our ditch, and thus conducting it below our claim.

We calculated that when the water was low, from July fifteenth to September fifteenth, a flume twelve feet wide and six feet deep would carry the entire river. It would require about two hundred and fifty yards of flume, and two hundred and seventy-five yards of ditch, twelve feet wide at the bottom, averaging thirty feet in depth; and nearly the entire distance through solid granite. It was an expensive undertaking, but we had no doubt that there was gold enough in the river bed to pay us well, could we only get it.

Companies for turning the river and working its bed, had their boundaries well defined, and to each its specific name; as Iowa claim, New York claim; ours was the Pioneer claim. There was also a definite understanding as to the joining of flumes and ditches.

The company next below ours tried to work their claim last year but failed; having raised a high dam, they were not able sufficiently to shut off the water. This year they intended to deepen their ditch, and thus lower the dam; and in accordance with this arrangement we planned the exit of our ditch.

Mr. Matthews, a shareholder, was employed as foreman, to direct the work, keep the record of labor, expense, etc. Aided by seven men he was to carry on the work, and as the river would not be low enough to turn out of its bed until the latter part of July, we had some eighty days in which to complete the job.

As there were fourteen shareholders each furnished a man every other week; and as McCord and I were camped together, we worked alternate weeks on the claim. This enabled us to spend half our time in mining, and still sustain our interest in the river claim.

Since enduring the fatigue, hunger and cold last winter, my health has not been firm, and on May twenty-first I was prostrated with rheumatic fever, and it was not until the eighteenth of June that I was able to resume work. However, McCord kindly took my place on the river claim, and when able to do full work I returned the favor.

The conflicting interests of river, bar, bank, and gulch claims, had long demanded some general rules of adjustment. For instance, it sometimes happened that a company would work a bar until the water of the river prevented their going further; but, when the river had been turned, so as to clear the bar of water, the company would return and demand the privilege of working out their claim. So also with gulches opening to the river; while those who, at large expense, had drained the river, naturally claimed all from which they had removed the water. Also the building of dams, causing claims to be flooded.

For these and many other matters, a meeting of mine owners, embracing the American river and its tributaries, was called to assemble at Mormon Island, on Monday, July 28th.

While there were delegates from all parts of the district, it was really a mass convention, as every mine owner was not only entitled to speak but to vote.

The meeting lasted but a single day, and yet rules for the regulation of all mining interests were read, discussed, adopted and registered. Many propositions were voted down, but I do not think there ever was a set of rules, which, upon trial, more perfectly bore the test, than those adopted by that mass convention.

Being the work of practical, earnest, upright men, they settled almost innumerable difficulties, and were recognized in the courts.

In this valley, the climate, affected by local conditions, is peculiar. April 16th it rained all day, and on the 17th there were a few showers. May 7th and 17th rained all day. June 10th two slight showers, the last rain until September 5th. Meanwhile vegetation dried up, and the surface of the ground became hard like brick.

During the summer months there seems to be a regular trade wind blowing up and down the river. After sunset the cool air of the mountains comes stealing down the river, to take the place of the heated air of the plain; the night becomes cold and one needs a heavy wrap for comfortable sleep. This continues until after sunrise, when there is generally a calm, and the temperature rises until after midday, often reaching one hundred or one hundred and twenty degrees in the shade, when a light breeze comes up the river, perhaps to take the place of the overheated air of the valley, but toward sunset it dies away.

To work in the sun during noonday heat is attended with great fatigue and danger of sun stroke.

In the early morning, before people get to work, the water of the river is cold and clear. Then we fill our buckets and place them in the shade. To a bucket of water we add about a pint of vinegar, and drinking freely of this, perspiration is promoted, and people work with safety, even in the hot sun.

Tuesday, June 24. For some time past, I noticed a man at work with a rocker on the bank of the river. His little tent stood near my path to the Pioneer claim, just above his place of work.

He was a fine looking man, industrious, unable to speak English, and as there seemed to be no one with whom he could intelligently converse, his utter loneliness impressed me.

Business led me out of my usual path for several days, and to-night as I returned by the place, the tent was gone, and there was a grave where it had stood. Inquiring of those who were tented in the vicinity, I was informed that some of them had noticed a stench proceeding from the tent, and upon examination the lone man was found dead in his bed.

Nothing indicating his name, friends, or country was found about him or the tent. There were no marks of violence on his person. He had evidently sickened and died alone; so wrapped in his cot he was buried, and his tent and effects burned.

It is sad indeed when sickness thus overtakes the stranger, and yet in these mines many die thus neglected and alone; not that people here are unwilling to help, but because the needs of the unfortunates are not known.

While there are many of the worst from all nations in California, I believe the mass of the people are equal or superior to any in the world, in intelligence, benevolence or courage. I saw this illustrated a few days since.

Two men were digging at the base of a sand bank, almost perpendicular and about one hundred feet high. Presently a large slice caved upon them, burying one completely, and the other to his shoulders. At the same time a large seam opened, extending to the top, showing that a great mass was about to fall. The one whose head was above the sand, called out, "Help here, quick."

About twenty men who were at work near by, made a rush to the spot; there was not a moment's delay, each brought his shovel, and there was perfect concert of action; not a second seemed to be lost, and the men were almost instantly released, when the great mass, like a mighty wave, swept down brushing the feet of the hindmost. It seemed, that had it fallen half a minute earlier, it might have been fatal to many.

## CHAPTER XII.

*River Turned from Its Bed—Good Prospect. Claim Flooded.—Resort to Law.—Sell Out.—Injure My Hand.—Rheumatic Fever.—Explorations.—Jesus Chico.—State Election.—Sickness of R. McCord.—Letters from Home.—Visit from Uncle John Steele.—We Visit Coloma.—Sutter's Saw Mill.—Peter F. Clark.—Return by Kanaka Bar.—Mining on Snyder's Bar.—Set Up Our Tents.—George Scott.—Snakes and Devils vs. Music and Angels.*

Friday, August 1st, 1851. All things being ready, this morning the river was turned into the flume, and gliding smoothly through its new channel left in the old bed only a series of ponds. Pumps were arranged at the lower extremity, where a low dam was thrown across to prevent the river from backing up at the mouth of our ditch, but it was not until noon the next day that we were ready to begin gold washing; and then a few buckets of gravel from the bed of the river, washed in a rocker, as a prospect, yielded over three hundred dollars worth of gold.

Elated with the prospect; hopeful that the reward of our labors was about to be realized, we went cheerfully to dinner. But on returning found the water like a placid lake, filling the channel, and our rocker, and such things as would float, on the surface; while crowbars, picks, etc., rested on the bottom.

The company below, finding the rock through which they had tried to cut their ditch very hard, had given up the task and again raised their dam. Reminding them of our mutual understanding, when, in April, we began work; they acknowledged their promise, but claimed that it would be very expensive cutting to such a depth through the flint-like rock; and besides, according to the rules made at the miner's convention, they had the right to raise the dam one more year.

Part of our company proposed taking up the flume, storing it until next summer and then set it up again.

Others argued that we would not only incur the delay, and the work of taking up and replacing the flume; but that another year we would not have the advantage of the ditch from which to receive the water into our flume; and consequently must build an extensive dam. These proposed an appeal to the law, in order to compel the removal of the dam which flooded our claim; and, being a majority, their plan was adopted.

Believing that the miner's rule permitting the raising of the dam would be recognized in court, I remarked, "Any one can have my interest, who will pay me fair wages for my work." The offer was immediately taken, by one of the company, who weighed out the gold, and my connection with the Pioneer claim ended.

It was fortunate for me, for after an expensive litigation, the miner's rule was recognized, and work on the Pioneer had to be abandoned for the time; and the next year there were difficulties about raising the water, so as to run it into the flume, and I never knew when the claim was worked.

Wednesday, Aug. 6. This morning while making an excavation on the bank of the river, my right hand was quite severely injured. However, coming back the next day with a rocker, I took out an ounce of gold, (sixteen dollars,) but my bruised hand became so painful I concluded to let it rest a few days.

Thursday, Aug. 28. On the ninth instant rheumatic fever again set in, and has troubled me occasionally ever since. But while not able to do much mining, I have made quite a thorough exploration of this region.

On either side of the river, along the foot hills, or terminating ridges of the Sierra Nevada, the soil is, no doubt, rich, though now barren, creviced and dry. It only lacks rain to make it productive; and some of the ravines are rich in gold, but cannot be worked successfully for lack of water.

To-day, while among the "rolling hills" on the Sacramento and Coloma road, I found a native Californian (Spaniard) vainly trying to manage a large herd of beef cattle from the coast region, intended for the mines.

For some cause he had been detained, and his herdsmen finding a saloon by the roadside, had stopped to await his coming, and when he arrived were all too drunk to render him any assistance; and the cattle, in search of water and grass, had scattered among the hills.

In broken English he told me his difficulty. Knowing of a marsh where the grass was still green, I mounted one of his herdsmen's horses, and assisted him to collect and drive his cattle to it. It was two or three miles from the road, but as grass and water were abundant, and the country around utterly devoid of vegetation, he could safely leave his stock, and spend the night at the Rolling Hills hotel.

He overwhelmed me with thanks, and when I refused pay for the few hour's service, he insisted on taking my name, saying his name was Jesus Chico, in English Jesus Little; inviting me, if I ever visited Santa Clara, to do him the favor of staying at his house. He was the first man I ever saw who was called Jesus, and the application of the name impressed me; afterward, through his acquaintance, I became familiar with the Spanish language, which somewhat shaped the tenor of my life.

Wednesday, September 3rd, 1851. The state election was held to-day. California was admitted into the Union as a state on the 9th of September, 1850, and though not yet a year old, great party spirit has been developed. In this locality there are four tickets, Whig, Democrat, Independent, and Miners and Settlers. Still in my teens and not yet old enough to vote, I could only look on as an interested spectator.

My health has so improved that tomorrow I expect to begin work in the bed of the river for the New York company. However, my illness has not compelled an entire loss of time; besides various explorations, I have read Mrs. Sigourney's Oriana, and the Legend of Oxford, Milton's Paradise Lost, Abbott's Young Christian, and Dana's Two Years Before the Mast; all very entertaining books.

Friday, Sept. 12. Since the sixth R. McCord has been very sick of fever. Leaving my work, I called a physician, and devoted myself, night and day to his care. He had been delirious until this morning he passed the crisis, his mind became clear, and though weak there are evidences of returning health.

Some time ago I made an arrangement with an express company to have my mail taken to, and brought from the post office at Sacramento, and to-day I was delighted to receive two letters; one from sister Loretta, the other from brother Edward. Bringing good news from a far country, they were indeed as cold waters to a thirsty soul.

By these letters I learned that an uncle, John Steele, after whom I was named, and who, with his brother William, left their home in Delaware county, New York, fifteen years ago, spending ten years in Georgia, from whence they removed to Green county, Missouri, where William died; and in 1850, uncle John crossed the plains, and was now in the vicinity of Coloma, scarcely twenty miles distant.

Sending a letter addressed to him at Coloma, I hopefully watched the express messenger for a reply.

Within a week McCord had sufficiently recovered to be left alone, I had resumed work, when, on Saturday, September 20th, while attending a miner's trial as a witness, my uncle having received my letter came to visit me. Some one pointed me out, and he advanced and took me by the hand. I had not seen him since I was four years old, but I felt less lonesome when he told me who he was.

While it is true that "there is a friend that sticketh closer than a brother," it is also true that the clanish spirit is natural, and however kindly others may regard us, the heart has a craving for kindred. It was pleasant to recall the various members of the family, and speak of the old and young from whom we seemed so far.

Monday, Sept. 22. Uncle John Steele and I made a journey to Coloma. Captain Sutter's saw mill stood silent and deserted; the bar, through which the race was dug in which gold was first discovered by Mr. James Marshall, in January, 1848, had not been molested, but the dam had been cut through, and the river bed and banks, for some distance above, were filled with busy miners.

Half a mile below Coloma the river curves around a long spur of the mountain; beneath this a tunnel, half a mile long, had been cut, and the south fork of the American river turned through, leaving the bed for a mile and a half quite dry. But the river bed was not as rich as anticipated; and I was informed that the company lost about forty thousand dollars in the enterprise.

Tuesday, Sept. 23. This is my first anniversary of entering the gold mines, and I am thankful that through sickness and suffering my life has been spared.

To-day I made the acquaintance of Mr. Peter F. Clark, a young man from Missouri, and friend of my uncle. Together we visited many of the mining operations along the river, and finally I concluded to join with Clark and my uncle in opening a placer mine on Snyder's bar, about three miles below Coloma.

Returning to Mormon Island the next day, I crossed the river at Kanaka bar, so named because it was occupied by several families of Sandwich Islanders, and English sailors who had married Kanaka women. Ascending the mountain on the south side, although the sky was cloudless, and the day warm, I had a very pleasant walk in the shadow of the great pines. When opposite Salmon Falls, again descending into the valley, I recrossed the river, and followed down its margin to Mormon Island. Found McCord much improved in health; arranged my affairs, and on the last of September returned to Snyder's bar.

Wednesday, October 1st, 1851. This morning my uncle, P. F. Clark and I set up a long tom, and commenced mining on the bar; but not being certain that it would pay, we built no cabin; did not even pitch a tent; simply fixed our camp under the shelter of some trees, a common mode of life here in summer.

For several days the weather was fine, and out door life very pleasant; but warned, by some light showers, that the rainy season was at hand, as our claim was paying pretty well, on the twelfth, we brought our tents and set them up, making a very comfortable dwelling. While here, we were relieved from the task of bread making, a baker and butcher bringing bread and meat daily, to the camps along the river, and as our mining was profitable we greatly enjoyed our stay on Snyder's bar.

Wednesday, October 21. George Scott, a young man, tented at the lower end of the bar, being intoxicated for some time, to day failed to appear; so in the afternoon my uncle went to his tent and found him suffering with delirium tremens.

He seemed intelligent, was fairly educated, and had but recently acquired the drink habit. After taking quite a fortune from the mines, while intoxicated, he gambled it away, and of course, when his money was gone, his companions deserted him.

We gave him some hot coffee and toast, and Clark and I watched with him during the night. Toward midnight his mind became clear; he was very weak, and for awhile seemed to be dying. Giving him a little more strong coffee, at last he sunk into a quiet sleep. This refreshed him, somewhat, and for breakfast took some more hot toast and coffee, but could not rid himself of the impression that snakes were crawling over him, and that devils were peering into the tent, ready to carry him away; and he plead with us so piteously not to leave him alone, that we took him to our tent, and my uncle remained with him.

Why is it that when people are reduced by drunkenness and debauchery that they feel snakes, and see devils? Are these their actual associates? and are they only discerned when the veil of flesh is ready to fall away? And how often when the pure in heart and life are brought low, they hear sweet music, and see angelic beings. Are not these their natural associations?

Scott remained with us several weeks, and so regained his health that he did good work; but became restless, and in spite of our efforts to persuade him to remain, went to Coloma and renewed his dissipated life.

The fate of George Scott has been the fate of thousands from Christian homes, who in the absence of home and church associations, have been tempted, and allured to their destruction by the drink habit.

## CHAPTER XIII.

*Last Meeting with McCord and Donnelly.—Downing's Ravine.—Placer Mining.—Pinoneros.—Indian War.—Uncle and P. F. Clark Return Home.—Start for Southern California.—Beautiful Scenery—Jesus Chico.—Pleasantly Entertained.—Journey Resumed.—Hernando Chico.—Tongues Loosened.—Doubted My Being an American.*

Wednesday, Nov. 5th, 1851. On the third instant I made a business trip to Mormon Island; was glad to meet my former partners McCord and Donnelly, with whom I had a very pleasant visit, and, returning to Snyder's bar, this morning bade them good bye.

We were more than common friends; brothers could not have been more devoted, and we confidently expected to meet often, but changing our residence, losing each other's address, the wild currents of active life drifted us strangely apart, and we never met again. Yet how often I have thought of them, and have always hoped that some happy chance might bring us together.

As the fall rains flooded the river we worked out our bar as soon as possible, and on the tenth began moving our effects to a place known as Downing's ravine, about five miles north-east of Coloma, into a cabin built by Clark and my uncle last fall.

The ascent from the river at Coloma was steep and difficult, but the place of our residence on the mountain was delightful. An open forest of pine, bur and live oaks, with an occasional clump of manzanita and chaparral covered the hills, and the ravines were rich in gold.

A small spring near our cabin supplied us with water for household use, and we trusted that the winter rains would furnish enough for gold washing. Here instead of granite, as along the river, the underlying rock is slate, and on this, mixed with gravel, in the bed of the ravines we found the gold.

This gravel is often cemented with a blue clay, which, taken freshly from the ground, is difficult to dissolve; like tallow, it seemed impervious to water, rolling into small balls, with the particles of gold adhering, thus carrying it away in the current.

However, we found that by heaping it on the bank where it could dry, it would dissolve in water like dust. Some of our best pay was from "tailings," which had been washed last winter; but then the clay had, unsuspected, gathered up the gold and carried it away.

We therefore devoted our time to digging and heaping up the gravel, where it might dry and disintegrate before washing. In the latter part of November there fell some copious rains, after which there was water for mining until March, in most of the ravines.

At the time of our arrival we seemed to be quite alone, but by the last of December were in the midst of a numerous population, and among our neighbors a large number from the gold mines of Georgia.

Here I noticed the work of the pinonero; a bird which picks holes in the bark of trees, generally pine, and then drives an acorn into each hole. Seeing the bark on the south side, seldom if ever on the north side of the trees, perforated with holes, nearly an inch in depth and diameter, we felt a curiosity to know what it meant; but it was all made clear when the acorns began to fall, and these birds were busy putting them into the holes. Thus removed from the ground, the acorns would neither grow nor decay, but furnished food for these provident little workers. All through the proper season they were constantly active, either making holes or bringing acorns.

White oak and bur oak acorns are much larger and of better flavor in California than in Wisconsin, and furnish the Indians with a large part of their food. They are often cooked and eaten with roasted grasshoppers. Not "locusts and wild honey," but grasshoppers and acorns.

About two miles from our cabin is a small village of Pah Ute Indians, known as Columbia ranch, which during the winter was involved in war with a village on the south, or opposite side of the American river near Placerville. After some weeks of indecisive warfare, each village, at the same time, sent out a war party with the evident intent of surprising the other; both made a detour eastward among the mountains, and met in a valley known as Rock creek, about four miles from the Columbia village, where a desperate battle took place, and a considerable number on either side were killed. The Columbia ranch was victorious, but among their dead was the oldest son of the chief Capitan Juan, (pronounced Cap-e-tan Whan.)

The Indians south of the river, trying to involve the Columbia ranch in trouble with the whites, sent over a small party which shot and killed a white man in Kelsey canon, near the Columbia village. But it so happened that a band of the Columbias were watching them, gave the alarm, and a party of whites pursued the murderers to their village, where the Columbia Indians pointed out the one who fired the fatal shot.

He was promptly arrested, but I believe for lack of proper evidence, was not executed. However, his village was greatly frightened, and fear of the whites virtually ended the war.

Early in the spring Peter F. Clark and my uncle returned to their homes in

Missouri, but wishing to see more of the country, I worked in the ravines until the water dried up, and then made a trip to Martinez, at the head of San Francisco bay, bought a horse and started out to explore the coast region.

My route lay between the Coast Range and the Pacific Ocean, and it was my intention to go as far south as Monterey, of which place I had read in Dana's "*Two Years Before the Mast.*" It was early in April; the rainy season was over; and the country was clothed in its greatest beauty.

Mount Diabalo, the highest peak in the Coast Range, was on my left, and as I crossed each spur, new scenes of grandeur and beauty came into view. The weather was all that could be desired; my lonely ride led through enchanted grounds, and it seemed that all I lacked was companionship to make the enjoyment perfect.

If Clark and my uncle had been with me, or McCord and Donnelly, or better than all, my brothers and sister, what a delightful journey it would have been!

But the balmy air, bright sun and every new charm of landscape, only added to my feeling of loneliness until I seemed to realize the force of the aphorism,

"The friendless owner of the world is poor."

Fortunately I found an American family with whom I stopped the first night, and late in the afternoon of the next day, called at a large adobe residence to inquire as to the possibility of finding a hotel, or English speaking family, or even the trail to Santa Clara.

While trying to make myself understood by a young Spaniard, who met me at the gate, a middle aged gentleman approached saying, "Meester Ste-le please, not will travel more to-day."

The voice and countenance seemed familiar, and, strangely enough, he recognized me as the one who came to his relief when his herdsmen were drunk, and his cattle were about to leave him at Rolling Hills on the Coloma road.

Leaving my horse in charge of the young man, Senor Don Jesus Chico very politely led me through a large gate into an open court, surrounded by a veranda, into a large room furnished with upholstered lounge and chairs, two small tables, and on the walls a variety of pictures. Here I was introduced to his family, consisting of Mrs. Chico, their two daughters Guadalupe and Jesucita, and their son Hernando. They were all very polite and kind, but only the father was able to converse in English, and his words and sentences were very broken. Now I felt the need of a knowledge of the Spanish language, and resolved to devote my time while in that region to its study and practice.

Senor Chico informed me that his son Hernando, and a number of herdsmen were about to go south on a business trip, and they would wait until I was rested and I could go with them and see the country.

It was just the opportunity I wanted, so I told him it would suit my convenience to start with his son in the morning, and at the time of my return he might expect me to converse with him in Spanish, and his son to join with us in talking English.

But he insisted that I should remain at least one day, and, pleased with my desire to learn Spanish, at once became my teacher, and soon put me in possession of quite a stock of Spanish words and phrases.

In the evening the daughters sung several Spanish songs, accompanied with the guitar, and insisted that I should sing something in English. I therefore sung a little ditty, named Hope.

"She comes our path to lighten,
To twine the diamond band;
Uniting earth and heaven,
That happy spirit land," &c.

The next day was devoted to talk for Senor Chico, or Mr. Little, as he liked to be addressed among his Spanish neighbors, and his son desired to hear me converse in English, and I was just as well pleased to follow them in pronouncing Spanish. In the meantime we visited the neighboring ranches, and the hamlet of Santa Clara, where I purchased a Spanish and English primer to assist in conversation with the people.

Wednesday, April 7th, 1852. This morning Don Chico proposed that I take one of his horses and leave mine to rest, but his kind offer was thankfully declined, and about nine in the forenoon our horses were ready. With a prayer that I might be under the protection of God, enjoy the journey, and an earnest invitation to return to their house, they bade me good bye, and I stepped to my horse, ready to mount. Hernando, kneeling before his father received his blessing; then after embracing his mother and sisters, mounted his horse, and with an affectionate "Adios," we rode away.

Native Californians are noted for their fast riding. Hernando was no exception, and soon we were out of sight of the Chico residence, but for some time neither was able to break the silence, and engage in conversation. So many things we wanted to talk about, but because we spoke different languages our lips were sealed. The silence was oppressive. At last I thought of my primer, and found, in Spanish and English, the question, What do you call this?—*Como se llama este?*

This was the key that unlocked our lips, and though sometimes it seemed like a vain repetition, our conversation never again lagged.

Before noon I had learned from him that four herdsmen had started early, and we would overtake them at the foot of a hill, by a spring among the trees; and he repeated the information to me, or rather after me, in English.

On many of the slopes there were large oak trees, which, instead of growing upright curved to the incline of the hillside, so that though they were fifty or sixty feet in length, a person on horseback could touch the topmost bough. At noon we descended into a ravine among large oak trees, and there at the spring we found the four herdsmen vaqueros. Having kindled a fire, they were preparing dinner, which consisted of bread made thin like pancakes; and then rolled up; cold boiled beans, well seasoned with red pepper, dried beef, and strong, hot coffee.

They also had wine, which they greatly praised, and seemed surprised because I refused to drink it. I learned afterward, that my refusal made them doubt my being an American.

## CHAPTER XIV.

*A Delightful Journey.—Salinas Valley.—City of the Angels.—Garden of Eden.—Return Journey.—Cuyama Creek.—Collecting Wild Cattle.—Herdsman's Skill.—Intelligence of the California Horse.—Return to the Chico Mansion.—Cordial Reception.—Among Native Californians.—Part with the Chico Family.—Return to Downing's Ravine.—Spanesh Speaking Indians.—Smallpox.—Seek Vaccination.*

For nearly two weeks we traveled through a most beautiful country. The houses were in clusters, and around them were many fine orchards and vineyards. Sometimes we passed through valleys which swarmed with cattle, and for two days Hernando and I, leaving the herdsmen, rode out of our direct route to obtain views of notable cliffs, canons, and of the sea. At two points we rode out upon cliffs which seemed to overhang the sea, where we could feel the rock tremble as the strong Pacific tide came rolling in.

Following up the Salinas river, we entered upon a mountainous region of wonderful beauty, and finally descended to *La Cindad de los Angeles*, (The city of the Angels.) About a mile north of the city we stopped with the family of Hernando's uncle named Jimnes.

Here we remained two days to let our horses rest, and I was glad of the opportunity, being almost tired out with the long ride. Owing to fatigue some objects of interest were not visited; but at the Jimnes home the orchards, vineyards, and flowers with the delightful climate, impressed one with the idea that here was the veritable Garden of Eden.

Leaving Los Angeles we traveled nearly due north, and the first day passed beyond the settlements. Some parts of the country had a desert appearance, and the distance was long between watering places. Heretofore we had stopped at a house for the night. Was always cordially received; but most of the way on our return journey we lived in camp and slept on the ground under the sheltering branches of trees; but the ground was dry, the air pleasant, and the stars looked brightly down from a clear sky.

At last we struck the headwaters of Cuyama creek, followed down half a day's journey, and then through the uplands on the right. Here the herdsmen spent several days in collecting a herd of beef cattle. When a few were brought together, they were driven into a small valley, where there was plenty of grass and water.

Here they were guarded until more were gathered, and then all were driven a good day's journey to some other place abounding in grass and water, where, being tired, they would remain very quietly. But just as soon as they were rested, or became uneasy, they were started again. Thus cattle were sometimes driven around in the same locality, simply to keep them from straying.

These cattle when first approached were generally very wild, and would sometimes scatter in all directions; but, unless chased, would soon come together again. If hemmed in, they would turn and fight, and then were exceedingly dangerous, it requiring great skill, and presence of mind in managing a horse when attacked.

However, our herdsmen understood all about it, and would so drop their riattas as to entangle and tame the most furious. And after an animal had been overthrown a few times, when approached by a horseman, he would generally stop and shake his head, as if expecting to be caught.

When the herd, which numbered over three hundred, had been collected and branded, the journey was resumed; and driving them before us, we descended into the valley of the Salinas, following down until we struck the trail by which we first came to the river.

As the herd was now accustomed to be driven, and tired enough not to wander nights, Hernando and I went on in advance, leaving the herdsmen to bring the cattle.

In that early day, when there were no fences, each animal in the ranchero's herd, horse, cattle, or sheep, was known by its brand, and it was the duty of his herdsmen to see that all the young stock was branded; and any one finding an animal of two years, or older, without a brand had the right to keep it.

Of course, unbranded stock were strays that had left the ranches when young; were generally found in the mountain region and were known as *ganados silcestres*, (wild cattle,) or simply *silvestres*, (wild ones.) It was these that Hernando and his herdsmen were in search of, and as fast as captured were branded, driven to his father's ranch, and ultimately to the mines and sold for beef.

On this trip I learned that the average California horse understood the movements and methods of the chase about as well as his rider. It was interesting to see him dodge the horns of a furious steer; how quick to notice when the riatta caught an animal, and place himself in a position to receive the strain, when it tightened on the pommel of the saddle with a shock that would often throw the steer headlong. Such exercise was very exhilerating, and with just enough danger to make it attractive.

My horse was an American, a dark chestnut Morgan; of fair, but not extra speed, spirited, but entirely unacquainted with the maneuvers of the herdsman; and therefore, several times I found myself at a disadvantage, and in dangerous positions, and appreciated, as at first I could not, the kindness of Senor Don Chico in offering me a well drilled horse.

On the third day after leaving the herd, we reached the Chico residence, and were welcomed by Senora Chico with "*Mil gracios a Dios por su venida sin dano.*" (A thousand thanks to God for your safe return.)

The greeting of all was so cordial, that they made me seem like one of the family. And although it was scarcely a month since I began the study and practice of the Spanish tongue, I found myself able, witq an occasional help from primer, to converse quite intelligently. In the meantime Hernando was making good progress in the practice of English. I could hardly call it a study.

But with me, I heard nothing but Spanish, only when my own words were echoed back by Hernando. Evidently the best method of learning to speak a foreign tongue is to hear, talk, and think in no other.

It was my intention to return at once to the mines, but Hernando insisted that my horse should rest a few days, while with another, I accompanied him on various excursions in the neighborhood, which he had planned.

Sometimes his sisters, Guadalupe and Jesucita, rode with us, and while they were polite and reserved toward me, ready to hear, quick to understand and reply in all seriousness to my conversation, yet, I felt sure, they were often merry, when by themselves, over my lame efforts to form sentences, and pronounce words in Castilian.

In thus traveling with Hernando, I saw many of the native Californians, visited their homes, observed their business methods, and was impressed with their lives of contentment and leisure. No one seemed to be in a hurry except when on horseback, and then they almost invariably moved at a sweeping gallop. But there was time to talk, and rest, and wait.

In business affairs they seemed to have adopted the maxim, "Never do to-day what can be put off until to-morrow." Around these quiet homes and drowsy hamlets there was the greatest possible contrast with the promptness, struggle and rush at the mines, where people could hardly find time to eat, rest or sleep; these Californians scarcely found time for anything else.

All classes seemed at home in the saddle; in fact it might be said of many that they spent their active life on horseback. The little babe when eight days old, is taken by the *padrinos*, (godfather and godmother) on horseback to the priest, when it is christened; and afterward, nearly every day, the child is carried somewhere on horseback, so that each one's earliest recollection is associated with the horse.

I believe that the native Californians (Spanish) were all devout members of the Roman Catholic church; paid special attention to its ceremonials, and reverenced the priesthood; but priests and people were addicted to drinking wine, made from the native grape; and a kind of brandy made from fruit; and drunkenness was sadly prevalent among all classes.

Another unfortunate habit was that of gambling, in a great variety of forms, and the strange thing about it was that none seemed to consider either drunkenness or gambling a vice.

No people could be more kind or hospitable. Politeness seemed natural; they never passed each other with indifference; and if one was about to shoot you he would probably first give you a most polite salutation.

For many years stock raising had been

about the only industry in this country; hence, the almost constant use of the horse.

A few years ago, ships from the United States and other countries came around Cape Horn to this coast for hides and tallow. In those days the flesh, having no commercial value, was thrown away. Since the discovery of gold, and the mines afforded a market, the meat only is salable, the hides and tallow being thrown away. This was in 1852, when cattle were slaughtered in the mines, and freights were too high to admit of the transportation of hides, tallow, and such things to the sea coast.

Thursday, May 6th, 1852. This morning I said *Adios* to the Chico family. At parting Senor Chico laying his hand upon my head solemnly invoked the Divine blessing, and that God would keep me in all my ways. Hernando traveled with me until near noon, when we took an affectionate leave of each other, tears filling his eyes as he said "Good bye, and God be with you."

Among the pleasantest memories of my life is my tour through the Spanish settlements of California, and my association with the Chico family.

At Martinez selling my horse to the same man from whom I had bought it, boarding a steamboat for Sacramento, where I purchased Ollendorff's new method of learning Spanish, also a reader and dictionary to assist in a proper study of that tongue, and, taking the stage for Coloma, sixty-four miles distant, in due time arrived at my cabin in Downing's ravine.

I now learned, what before I had not even suspected, that many of the Indians were familiar with the Spanish language. Widely dispersed throughout the country, were those who, a few years ago, were in employ of Captain John A. Sutter; some as laborers, and many others as drilled and disciplined soldiers at his fort.

But the change in the government, the inflow of immigration, and the building of Sacramento city, had broken up the old order, dispersed the soldiers and laborers, and while they still retained their native dialect, the Spanish language, in which they had been trained, was not forgotten.

On reaching Downing's ravine and learning that smallpox was prevalent among the miners, and fearing exposure and attack I immediately returned to Sacramento for the purpose of being vaccinated, and remained until it became effective.

## CHAPTER XV.

*Smallpox at Indian Village.—Vaccination.—Pah Ute Funeral.—Emblems of Mourning.—Go South with John Ford.—Trouble at Moqualomne Hill.—Mystery of "The Gentleman from Mississippi."—Joaquin.—Desperate Leap for Life.—Under Suspicion.—Danger of Being Lynched.*

On my return, meeting with Captain Juan, chief of the Columbia village, he told me in Spanish, with which I found that most of them were familiar, that one of his people had died of smallpox, and others were sick. Explaining how a person, by vaccination, could escape, I showed him my arm, telling him I had no fear, for, after a person was vaccinated, smallpox would not make him very sick, and, taking some of the virus from my arm, vacciuated the chief, and his son who happened to be with him.

Having in my cabin a hawk's wing, I took a quill, and filling it with the virus from my arm, went with him to the village and vaccinated quite a number, showed them how, and advised them to vaccinate every one, old and young.

These Indians burn their dead. A pile, usually of dry manzanita about six feet long, three high and three wide, is prepared and the body neatly rolled in a blanket, or other clothing is laid thereon. Fire is then applied. The people form a circle around it, and led by a master of ceremonies, engage in a mournful chant, or dirge.

Respect for the dead is indicated by the value of the offerings placed on the fire.

When the body is reduced to a cinder, it is taken out of the fire, folded in a cloth, and sometimes wrapped with strings of beads. When it is properly prepared by the leader, it is passed from hand to hand around the circle, each one upon its reception, turning away from the fire and holding it up at arm's length, says reverently, "To Thee O God."

When it has gone around the circle, the embers and coals are raked together, and the master of ceremonies again commits it to the fire; and when burned to ashes they are taken up, and the nearest relatives use them to paint black lines upon their faces. These are the emblems of mourning; and the form in which the paint is put on, indicates the relationship of the mourner to the deceased.

Aware that the entire village would soon feel the effects of the vaccination, and fearful that they might think that I intended to kill them all, it seemed to me prudent to keep out of their way for awhile. So in company with John Ford, a young man from Georgia, I made a trip to San Andres, exploring the Moqualomne, Calaveras, and Stanishlaus rivers.

In California the different nationalities did not always harmonize. Those of different speech, not being able to understand each other, sometimes had serious quarrels.

Such we found to be the condition at the mining town of Moqualomne Hill. Reaching the place about dark, after supper we walked through the village to converse with the miners who had come in from their work. Passing into a large store, which seemed thronged, we were addressed in Spanish, to which I replied in the same tongue. Mr. Ford made some inquiry in English, when I heard some one exclaim with an oath, "*Es Americano, matele! matele!* (He is an American, kill him! kill him!)

At first I doubted my understanding of the words; but when a knife was flourished, and a rush made at Ford, knowing there was no mistake, I grasped the arm which held the knife, as it came down; and yet, in trying to parry the blow, Ford had his right hand severely cut. With a bound we were out of the store; and, utterly bewildered at the unprovoked attack, lost no time in reaching our hotel.

The landlord informed us that for some time a bitter feud, about some mining claims, had existed between the Spanish and English speaking people; that they lived in separate parts of the town; and this afternoon there had been a collision, several shots had been fired, and probably some one had been killed.

In our ignorance of the conditions, we had wandered over into the Spanish end of the town, and hence the clash. Ford's wound was dressed, and early the next morning we left the warlike camp.

At San Andres I found the solution of a puzzle which had been presented in Downing's ravine.

One night, shortly before starting south, one of my neighbors, a gentleman from Georgia, brought to my cabin a fine looking man, whom he introduced as Colonel Davis, brother of the senator from Mississippi, who was making the tour of California, and would like to stop with me in my cabin a few days.

It was not often that I was honored with a guest possessing such distinguished affiliations, and therefore did my best to make his stay pleasant. Telling him of my plan to visit some of the southern mines I expressed the thought, that if he had not already been there, we might go together as far as Jackson or San Andres.

"Have you friends there?" he inquired. I thought I had, if I could only find them. In fact I found two at Volcano, James and William S. Hanford, of Walton, New York.

The next morning I apologized for our plain fare, but we hoped to have something better for dinner, and he said I might expect him precisely at noon. Noon came, dinner was ready, but the Colonel failed to appear, and I never saw him again; nor could I find any one who had seen him after he left my cabin.

My Georgia neighbor, who introduced him, knew not where he had gone. I was greatly puzzled, and feared that his friends might suspect me of being his murderer. But after reaching San Andres the mystery was cleared up.

A man named Davis, answering exactly to the Colonel's description, had lived at San Andres, but whether really any relation to senator Jefferson Davis might be doubted, although there was a resemblance in person and features.

While under the influence of liquor, he went into a barber's shop kept by a negro, whose family occupied part of the same house. Going into the family room he insulted the barber's wife, and was ordered out. Not being inclined to go, the barber came to protect his wife, and very properly demanded. "Leave here, Sir, or I'll kick you out."

The so called Colonel deliberately turned to him and said, "No *white* man ever talked that way to me and lived;" and, presenting a pistol, shot the barber dead in the presence of his family.

The murderer, pursued by the indignant citizens, tried to make his escape, but in crossing a ravine, filled with washings from the mines above, sunk in the mire and was captured. He was neither shot, hanged, nor burnt; but was handed over to the sheriff of Amador county, and lodged in the jail at Jackson.

Of course there was great indignation against the murderer, and a few nights after his arrest a crowd appeared before the jail and demanded the prisoner. The sheriff, supposing they intended to kill him, made the jail as secure as possible, and tried to persuade them to let the law take its course.

In the meantime, another small party came, privately assuring the sheriff that it would be impossible to keep the mob out, and the only way to save the prisoner and honor the law, was to place him quietly in their hands, permitting them, without the knowledge of any one in or about the jail, to remove him to another place, and after the mob had searched in vain, of course the sheriff would be honored for the wisdom of his strategy.

The plan was adopted, the prisoner delivered up, and under cover of the night, conveyed to a place of safety. After a little more parley, the sheriff informed the mob that the prisoner had been removed and was entirely beyond their reach, and to verify his words, invited them to come in and search the jail. Not finding him they concluded that the sheriff was the right man in the right place, and the public interests were safe in his hands.

But the sheriff never again saw his prisoner, he having been placed in the hands of his friends and associates, who not only wanted to get him away from the mob, but out of the hands of the sheriff, and they succeeded. Just how the sheriff settled with the county I am not certain, but was told that he claimed his prisoner was finally taken by a mob, and whether put to death he did not know.

He was evidently the same man who stopped with me in Downing's ravine, and his friend, my Georgian neighbor, was helping him to escape justice. No wonder he became alarmed, and skipped out, at the mention of San Andres. Truly, "The wicked flee, when no man pursueth."

About this time, a Mexican named Joaquin, a notorious desperado, and leader of a gang, who, by murder and robbery, were a terror to the country, had been traced to the neighborhood of San Andres. One evening, while at supper in a hotel, he, being unknown to any about the place, seated himself at the supper table.

Back of him was an open window, and some twenty feet below was a water ditch probably ten feet wide, and on the opposite side piles of broken rock. He faced the door and windows, which opened upon the street, and as I sat nearly opposite to him at the table, my back was toward the door. He was a fine looking man, and I had no idea who he was, but judged from his appearance that he was a Mexican, and wishing to improve every opportunity to practice my newly acquired Spanish, gave him the usual salutation, "*Como le va Senor?*" (How do you, Sir?")

"*Muy bien,? De donde V?*" Very well, where are you from?)

"*Del norte, cerca de Coloma.*" (From the north, near Coloma.)

As neither he nor any of his gang had operated in that region, he was evidently sure that I had no suspicion as to who he was, and so the conversation ran on.

Suddenly he arose, turned to the window, and as several shots were fired, sprang out. Whether he was hit, I do not know, but it was a desperate jump across the ditch on those rocks; and although it was hardly dark, he disappeared in a large growth of chaparral just beyond and made his escape. The sheriff's posse had surrounded the house except on that side, not thinking it possible that any one could pass in safety from that window.

Seated with my back toward the entrance, I had not seen the attacking party; but there were those who had observed me in conversation with Joaquin, and under suspicion, I was held until the pursuers returned, and then put through a rigid examination.

Mr. Ford explained whence, how and when, I came to San Andres; but his wounded hand excited distrust, and for awhile both were in serious danger; not from the sheriff and his posse, they were satisfied with our innocence, but from the unreasoning crowd, insisting that we belonged to Joaquin's gang, and of course ought to be lynched. I am sure that one who has never faced such a condition, can have no idea of the situation.

However, we were both young; certainly not hardened criminals; and as I could refer to well known men in Coloma and Sacramento, we were at last entirely relieved from suspicion.

A large reward was offered for the capture of Joaquin, dead or alive, and a year or two after this, he was killed by a sheriff in trying to effect his arrest.

## CHAPTER XVI.

*Return to My Cabin.—Distinguished Reception.—Gift from Captain Juan.—Elijah Barker.—One of the Saints.—Spanish Flat.—American Flat.—The Fraudnlent Sale.—Muster for a Fight.—The Advance. —Terrible Danger.—Final Explanation.*

Two weeks from my departure south I returned *to my cabin*, and was surprised to find myself regarded by the Columbia Indians as a great medicine man. Most of those taken with small pox had died, but after vaccination there were no new cases; and it was, no doubt, well for them that they burned their dead; thus, with them, consuming their infected clothing.

The Chief, Capt. Juan, accompanied by his son, and principal men of the village, made me a formal visit, thanking me for the benefits conferred in vaccination, and asked whether there was anything they could do for me. In reply I told them that it was my wish that we might be friends, and that they would treat me as a brother.

The Chief carried a beautiful cedar bow, along the convex side of which, as neatly as the bark on a hickory sapling, was fastened the sinew from a deer's leg. He had also twenty-five feathered and flint-pointed arrows, in a quiver resembling a fox skin, only the hair was black.

After an examination, I inquired, "Would you sell them?" He said, "No, I will give them to you." Then calling the attention of those present, he said, "*Este arco, estas flechas con esta piel de jau, pertenece a mi joven hermano blanco.*" (This bow, these arrows, with this fox skin, belongs to my young white brother,) and rising up he placed them in my hands.

I was greatly pleased with the gift, and, with sincere thanks, assured him that they would always remind me of Capitan Juan and his people. I kept them with greatest care; and when on my way to New York, had them in a neat box, but while going down the San Juan river, in Central America, they were stolen from the boat.

Among the first with whom I became acquainted in the vicinity of Downing's ravine, was Elijah Barker, a colored man about forty years old, a slave, whose owner, James Barker, had brought from Georgia. Peter F. Clark and my uncle had known him nearly a year longer than I, and spoke of him as an excellent Christian man.

His master, generally known as "Jeems" Barker, had the reputation of being unsteady; after reaching the mines was soon out of money; but Elijah was hired out, and when he had earned enough money, "Jeems" concluded to go back to Georgia, where expenses would be less. He would have taken Elijah with him but for lack of means. However, Elijah was at work, and doubtless when some of "Jeems" Georgia neighbors were ready to return, he would have earned enough to pay his fare and might go with them.

But he discovered a mine, and working on his own account, was soon in possession of considerable gold. Very industrious, he worked in his mine during the day, and often in the evenings washed clothes for the miners.

My uncle had read to him the letters sent by his master, answered them, and assisted him in business matters; and after he left Elijah came to me for such help, and so by reading and writing his letters, assisting in his business, I became familiar with all his affairs. He was intelligent and sociable; relating many incidents, some humorous, others exceedingly sad; all of which gave me an inside view of slavery.

Slaves took the sir name of their master, and he, by being sold, had his name changed three times, and finally, being given as a dower to James Barker's wife, took the name of Barker. He was married, had two children, but his wife belonged to a man named Grove. He often spoke of them, and always sent them an affectionate message in the letters addressed to his master.

He was much worried, fearing the Grove estate might be sold, in which event he might never see his wife and children again, and sometimes when expressing these fears, he would break down and weep bitterly.

Of course I was interested. Grove had come to California, and learning that he wanted to sell his slaves, and bring his family, I suggested to Elijah that he buy his wife and children, and have them come with Grove's family to California, where they would not only all be together, but free, because slavery was not recognized in California.

He replied, "Yes, Massa John, Ize thought 'bout dat; but it can't be done." And then he sobbed as though his heart would break.

"O, yes," said I, "I'll transact the business for you, and you need not pay a cent until they are here; and if you lack means I'll make it up, and trust you to make it good."

Still he objected, but always, when we met, the conversation turned upon that subject.

At last a letter came from his master requesting Elijah to return with certain Georgians who were about to leave California.

When they were ready to start, he came to bid me good bye; and I made my final appeal, urging him to rescue his wife and children, and showing how happy they all could live together in Cal fornia. It was evidently his greatest desire; but instead of acquiescence, he utterly broke down, and wept for a long time.

At last, with a great effort, overcoming his emotion, he wiped away his tears, and rising up, said with deep solemnity, "Massa John, de Lord heard me promise massa Jeems dat I'd come back, an ob cose I will."

Nothing could tempt him to break his word. From that time he seemed to me like one of the old saints or martyrs. All his life a slave and yet so near to God. As surely as that the fear of the Lord is the beginning of wisdom he was wise. In the presence of such faithfulness I felt humbled.

With an earnest prayer for my salvation, prosperity and happiness, bidding me good bye he started for Georgia and slavery. But he died on the way; massa Jeems obtained his earnings, his wife and children were sold with the Grove estate; and yet, it is possible that they are all together in a better home than all the wealth of California could furnish.

In the early history of the mines a company of Mexicans occupied a prairie-like slope about three miles from my cabin; and were joined by some native Californians and Chilenos, and, as they all spoke Spanish, the place was known as Spanish Flat. Westward about a mile through dense pines, was a similar place, occupied by a company of American miners, and hence, called the American Flat.

About one hundred Chilenos arrived in San Francisco, and coming out to the mines, very naturally came to those who spoke their own tongue; and so, in the latter part of June, 1852, Spanish Flat became quite populous.

One Sunday, when the American company, only five or six in number, were away, some persons, claiming ownership, sold the American mine to the newly arrived Chilenos, and receiving the price, a considerable sum, left before the fraud was discovered, and they identified. Having bought not only the mine, but mining implements, the Chilenos immediately began work, and when Messrs. Burt and Grove, the real owners, came Monday morning, they found that their claim had been "jumped." Probably a hundred men were in possession, and ready to hold it by force of arms; and as they spoke different languages, explanation was impossible.

Messengers were sent to the nearest mining camps, asking the men to bring their rifles and other weapons, and to assemble on the ridge above the American Flat. At that time, June 7th, 1852, most of the miners had left the uplands for the rivers; and by three in the afternoon, rifle in hand I reached the rendezvous, where only some forty had assembled.

A man named Murphey explained the affair as he understood it; knowing nothing of the sale, but stating, in effect, that probably one hundred or more Chilenos on Sunday had taken forcible possession of Messrs. Burt and Grove's claim and tools, and refusing to give them up, threatened to shoot whoever interfered. Of course, such a force would soon work out the mine; and he proposed that the miners present, drive away the Chilenos robbers, shooting them down if necessary, take possession of the sluices before they were cleaned up, and the gold panned out, and restore the property to Burt and Grove whom we all knew to be the rightful owners.

All agreed to this, and after some preliminary drill, and the understanding that in the event of a fight, we must stand by each other, the little company left the timber and marched for the mine. It was only about forty rods from our place of meeting, and at our appearance, the Chilenos, dropping their implements, took up their guns and immediately formed a line of defense.

They were, no doubt, expecting us,

and had been reenforced from the Spanish Flat. Without the least parley, or chance of peaceful agreement, our leader seemed about to precipitate a bloody conflict. But at his command when within about fifteen or twenty rods of their front, we gave a yell, made a rush, at the same time raising our rifles as if ready to fire. Just then a few left the center of their line, taking shelter behind a bank of earth, and the next moment the whole body was in confusion, and rushing for the timber in the direction of Spanish Flat.

Fortunately not a gun had been fired on either side. We followed them, but they kept well in the advance, and as we came out of the timber, a native Californian met us, gesticulating and shouting, "*Dos horas! dos horas!*" (Two hours! two hours!) Soon it was understood that the Chilenos wanted two hours in which to get readv to leave; but our leader gave them to understand that if they were not gone in one hour, they might expect to be fired on.

Many of our company considering them robbers and dangerous characters were willing to shoot them down; and prob ably within half an hour the Chilenos who had participated in working the Burt and Grove claim had all gone.

In the meantime I had an interview with the Californian, and he related the story of the sale; said it was made in good faith, and that he, with others present, would be able to recognize those who had sold the claim and received the money.

When I told Murphy, our leader, he insisted that they should come and see whether those who made the fraudulent sale were in our company; but after an examination they decided that they were not.

However, they so described them that some of our company recognized them as two men who had followed gambling in this vicinity. The villains were not found; but the unfortunate Chilenos lost their money, and were driven like criminals from the community.

I was decidedly ashamed of my participation in the affair, and lost no opportunity of explaining to both Spanish and English speaking people the perfect innocence of the Chilenos.

But it was a terrible danger to all concerned; and the Californian, who was with them, said, had it not been that they were far from home, and surrounded by a people with whom they could not converse, the Chilenos would have stood their ground, and shot us down, for they believed we were simply a band of robbers after their property.

## CHAPTER XVII.

*Texas Bar, South Fork American River.—How Placerville Became Hangtown — Miners Murdered.—Intimidation.—Desperados on the Street.—The Fatal Shot.—A Young Life Lost.—Pleasant Associations.—Personnel of Our Mess.—Protectection against the Pulex Irritans —Indian Camp.—Indian Mothers Teaching their Children to Swim.*

Early in June, 1852, Thomas Finney, John Stevenson, John Van Benschoten and I, organized a mining company to operate on Texas bar on the south fork of the American river, and when the river had fallen to the proper stage, we commenced work; putting in a very profitable summer.

It was only about two miles from Placerville, then generally known as Hangtown, from the fact that at that place five men had been hung the same day on one tree. As related to me by one who professed to have been an eye witness. There was in the village a saloon and gambling den known as the headquarters of a notorious gang of thugs. Men supposed to have gold, were killed and robbed on the streets at night; others were murdered in their cabins.

No one felt safe, either on the street or at his work; and yet no one doubted as to who were the criminals. Finally at the funeral of a man whose murder and robbery had been traced to certain gamblers, the citizens resolved that every professional gambler, that is, every one who followed no other occupation, must leave town within twenty-four hours.

At this they became more bitter and defiant than ever; and a leading citizen, who had obtained evidence involving four of them in the murder, was shot and killed while passing their headquarters.

The enraged miners immediately gathered, surrounded the saloon, and finding the four against whom charges of murder had been made, and who were suspected of firing the recent fatal shots, seizing and binding them, hands and feet, without further ceremony hung them to a tree.

The proprietor of the saloon became very angry, charged the crowd with murder, and threatened to avenge the death of his patrons. The mob was in no mood to listen to such talk; there were those present whose friends had been shot down and robbed, as they believed by those men, and suspecting that he was an accomplice in their crimes, a rope was quietly obtained, prepared, and suddenly the noose was slipped over his head, and he was dragged to the tree, and hung up with the rest.

Although the twenty-four hours were not yet expired, the mob concluded to finish the business, and get rid of the professional gamblers. But upon making search not one could be found; they had taken the hint and gone.

This affair occurred more than a year before we began work on Texas bar, but we found that many of the roughs had returned, or others had come, and quarrels and shootings were of frequent occurrence; but so long as they were confined to the saloon element, the citizens paid little attention to them.

Placerville was the base of supplies for a large mining region. As in other mining camps, so on Texas bar, provisions and mining implements were paid for by the miners, and distributed by the merchants. And so it often happened that miners going to Placerville, bought not only for their own, but for other companies, and thus often carried large quantities of gold.

Mr. Anderson, one of these agents, from Chili bar, half a mile down the river from our camp, on his way to Placerville, in the early part of July, 1852, was murdered and robbed in Placer canon. A week later, another miner met a similar fate near Placerville.

These murders created a profound sensation, and while there was a lack of positive evidence, there were strong suspicions of guilt resting on certain persons; and in many camps the question was discussed of making another raid on the saloons and gamblers, but many people had been intimidated and feared their enmity.

As a matter of fact, those who were recently killed had given offense by publicly suggesting the suppression of the thug element; and, with others, had been threatened with violence. But while people generally would feel safer to have them suppressed, there was a common hesitation about opposing them, and individuals shrunk from becoming targets for their wrath.

While miners, merchants and mechanics had real interests at stake, those dreaded enemies of society, after the commission of crime, need only run and hide themselves in some other locality. None better understood the force of bravado and bluff; and busy people, while engaged in their business pursuits, were often greatly annoyed by them, though sometimes their reckless interference brought them to grief. I give an example.

One afternoon in the latter part of August, 1852, I went to Placerville for supplies. After completing my purchases, and was ready to return, I discovered two desperate characters on horseback parading the streets. Both were, or pretended to be intoxicated, and flourishing large revolvers, rode furiously while shouting to people on the streets, "Hunt your holes! hunt your holes!" And of course, people tried to keep out of their way.

Hoping to avoid an encounter with them, I remained some time in the store; but supposing they had gone, at last ventured out. Placerville at that time was composed of two clusters of houses, with quite a space between. Scarcely had I left the shelter of the store, when here they came, over a little ridge from the upper town; one on each side walk, flourishing their pistols and howling at the top of their voices.

A large, well proportioned man walked a short distance in advance of me; clay-stained clothing indicated that he was a miner, a coat lay on his left arm, attached to his belt a large revolver hung at his back; and on the seat of his pantaloons was a large patch, evidently a piece from a flour sack, as it bore the mark EXTRA FINE.

He would first meet the reckless rider, and I hesitated to see what would be the result.

Nearer came the man on horseback, still flourishing his pistol and shouting, "Clear the track! clear the track!"

A shot from the horseman's pistol glanced along the side walk. The miner's hand had been laid upon his pistol; now it was instantly drawn and fired.

The rider threw up his arms; then made an effort to grasp the saddle, but fell heavily to the side walk; the horse shied into the middle of the street, the rider on the opposite side went quietly down to the South Fork, a noted gambling headquarters. The fate of his comrade seemed to have tamed or sobered him.

When I reached the body, the miner stood beside his victim; with some emotion he said, "I'm powerful sorry I had to do it; but I won't be shot nor run over if I can help it."

Thinking he might have been stunned by the fall, we tried to raise him up; but his body was limp and lifeless, the blood flowing profusely from a wound in his left breast. I don't think he regained consciousness.

He was a fine looking young man. His life might have been of inexpressible value to himself, and an honor and blessing to his friends. Alas, alas, for the use he made of it! and then vainly threw it away. Surely, "he died as the fool dieth."

A crowd gathered and the body was carried away. That the man who slew him did it in self-defense was not questioned, and the event soon ceased to elicit remark; but doubtless he was remembered somewhere.

Our associations on Texas bar were very pleasant. There were several companies, and among them several graduates from eastern colleges, two of whom had made the tour of Europe and Palestine, while our own mess was very congenial.

Mr. Thomas Finney from McHenry county, Illinois, some fifty years of age, an intelligent Christian gentleman. Threatened with pulmonary consumption, in 1850 he crossed the plains in search of health; found it before he reached California, and about two years residence seemed to have given him perfect soundness of body. He was a diligent Bible student, respected and beloved by all who knew him.

John Stevenson, some thirty years of age, a native of England, but for some years a resident of Boon county, Illinois; intelligent, genial, unselfish; of whom it was said, "He never acquired a bad habit."

John Van Benschoten, a native of Delaware county, New York; some twenty years of age, rather reticent, with a mind for business methods, and of untiring energy.

He was one of the victims of the New York shipping swindle, mentioned in

Chapter X, but was fortunate in being able to make the trip from Panama to San Francisco by steamer.

In the mines we all worked alike, but outside of this, each sustained his relation to the others according to taste or adaptation. For instance, Mr. Finney was our adviser, the Nestor of the camp; his wise counsel and sweet spirit exerted an influence for good to all on the bar. Mr. Van Benschoten was general business manager; Mr. Stevenson was cook, and your humble servant, the author, assistant.

Our habitation was simply some posts placed upright in the ground, supporting a roof of boughs. It made a good shade, and as there was no rain in summer, we enjoyed the out door air. A tent was pitched beside our booth, but was seldom used.

At first we were greatly annoyed by fleas, (pulex irritans) they seemed to be everywhere, but more especially where we fixed our bunks and tried to sleep.

However, we were advised to place our cots on a pile of the branches of a kind of black alder, which fleas avoid, and in doing so, were much relieved.

In the latter part of August a band of forty or fifty Indians camped on the opposite bank of the river, spending about two weeks mining and fishing.

Just below Texas bar, the stream descended in a narrow, swift channel, among large granite boulders. Here, with long spears, they caught many fine salmon.

All these Indians, even young children, seemed to be expert swimmers. And no wonder, for thev were compelled to learn, while scarcely more than infants.

Often several mothers would take their small children to the top of the rapids, where one after another would be dropped in, and we could see their little black heads in the white foam, bobbing around the great boulders, in the swift current, until they reached the eddy below, where there were always several ready to take them ashore.

They did not always want to go in, but the mother would say, "Shut your mouth," and drop them in. It was frightful to see them swept down through the deep water. Sometimes there were indications of strangling, but generally, when taken out, they would shout and laugh, as though they had not only done some great thing, but enjoyed it.

## CHAPTER XVIII.

*Peg Leg Smith.—Amputated his Own Leg. —His Daughter.—Arrival of Uriah and Charles T.—P. O. Soper.—Money vs. Kinship.—Winter Rains.—Alone in Camp.—Find a Partner.—Start for Downing's Ravine.—A Memorable Night. —Soper's Sickness.—Trouble with Mr. T. —Bullet through my Hat.—Suspicions. —Relief of Mr. A.*

While on Texas bar I became acquainted with Mr. Smith, who then lived at the little hamlet of Kelsey, on the bluffs about three miles north of our bar. He had been a mountaineer and scout; had married a squaw, who dying had left him with one child, a girl, at that time, 1852, about ten years old.

He was familiarly known as Peg Leg Smith, and had rendered himself famous by having amputated his own leg.

In a difficulty with some Indians, one fired at him, shattering the bone just below his knee. Aware, that without amputation, the wound would result in death, and as none of his companions were willing to undertake the operation, he resolved to do it himself.

Therefore preparing bandages as best he could, having a fire, in which after removing it from the stock, he heated the barrel of a horseman's pistol, with which to sear the ends of small blood vessels; with improvised grip, and threads to bind the larger ones; and sharpening his knives to a keen edge, with his own hands he severed his leg at the knee, and with the help of a comrade, who had not the nerve to undertake the cutting, bound it up in good shape; and years afterward, when I knew him, he was able to walk quite actively on a wooden leg of his own manufacture.

His daughter, lively, intelligent and shy, possessed fine features, and considering her Indian blood, and habit of going barehead in the sun, was strangely fair, and might be called a pretty brunette.

About the first of September immigrants from across the plains began to arrive at Placerville, the direct route for those who came by way of Carson's river, and two young men from Michigan, Uriah and Charles T.——, brothers, came to Texas bar and began mining, but finding the place would not pay, removed to another on the bar, where they were more successful.

Two weeks later their cousin, Mr. P. O. Soper, arrived from Indiana, and his two cousins, taking advantage of the confidence naturally arising from kinship, sold him their rejected claim. He was just recovering from a severe illness, and with more ambition than strength, a week's faithful effort not only proved the mine worthless, but nearly wore him out.

However, he gave them all the gold he had taken out, and asked them to cancel the bargain. This they refused, insisting that the contract should stand good, whether he could get the amount out of the claim or not, and if he would not agree to this he should not remain in their camp another night.

Having paid for all the provisions he had used, and learned from others on the bar, that, after testing it, they had sold him a worthless claim, of course he was willing to leave them.

Saturday, October 2nd, 1852. The winter rains had set in, the river had risen somewhat; my three partners had gone to the uplands to prepare for winter diggings; having pitched my tent I was working alone on a corner of our bar, which still paid fair wages, but was liable to be submerged at any time.

Rain had fallen all day, and a little before dark, going up on the hillside for fire wood, I met Mr. Soper, and supposing he was on a similar errand, in a familiar way suggested that while rain was good for green things generally, it was doubtful whether it would benefit him.

After some pleasantries, he told me of the trouble with his cousins, and that he was now on his way to a place about five miles beyond Placerville, hoping to find a friend with whom he had crossed the plains.

I knew the road; difficult in daylight, but in the coming darkness and rain it seemed impossible, especially for one so broken in health, and inviting him to my tent, advised him to wait, at least, until morning. After some hesitation the invitation was accepted.

His work had been too hard; he was prostrated for over a week, but finally recovered sufficiently to assist in my work. He was very companionable; well educated; English being his native tongue, but read and spoke the German; and was a very fine singer. We soon became attached to each other, and I invited him to spend the winter with me at Downing's ravine.

It was November before we left Texas bar, but most of our effects had been removed, and we waited a few days for our tent to dry, so it could be taken with us. The rains however continued, our mine was flooded; at last there appeared a break in the clouds, and crossing the river at Chili bar we started for Downing's ravine.

Delayed at the ferry, it was late in the afternoon when we began the ascent, and quite dark when we reached Dutch creek, where, to our dismay the great pine chopped across the channel, and forming a footbridge had been swept away.

I remembered another, about a mile

further up, but our way lay across a flat perforated with mining holes, twenty or thirty feet deep, and now all nearly full of water, and an approach to the soft edge of one of these would be exceedingly dangerous. It was a night of Egyptian darkness; the clouds had returned; rain fell in torrents, and groping our way to the trunk of a large pine we took shelter on its lee side.

Our clothes wet through, matches too damp to kindle a fire, the cold constantly increasing until the rain changed to snow, and in the piercing north wind it seemed as though we would perish; but could only sit there and shiver, as hour after hour passed away.

When the dawn enabled us to distinguish the shaft's dark mouth from the white snow, we hurried on, and in due time reached our cabin. But while life lasts we will remember that night on Dutch flat.

However, my only inconvenience, a severe cold, of which a good sweat relieved me; but Soper was again prostrated, and it was several weeks before he could engage in work. For some time I feared he would die, and especially regretted that there was no physician within reach; but applying snch remedies as could be obtained, and watching with him day and night, with all possible care; at last he rallied, and came slowly back to health.

When able to care for himself during the day, I improved the rains in washing gold along the upland gulches; always profitable when there was plenty of water.

Returning to the cabin one afternoon, I found one of Soper's cousins, who had come to demand pay for the worthless claim which they had sold him. He was inclined to doubt Soper's statement that even though willing to pay, he did not then have the gold. Perhaps not aware of his sickness, but supposing we had been doing a lucrative business, he seemed resolved to compel payment of his unjust demand.

Of course it was no affair of mine, and I intended to say nothing about it, but after considerable abusive language and threats, holding his rifle in his left hand, with the right he gave Soper a blow on the face.

Snatching my pistol, and holding it behind me, out of his sight, I said, "Hold on T——, that will not do here."

Turning his rifle toward me, and bringing his right hand to the lock, ready to raise the hammer, he replied, "Do you take it up?"

The next instant my pistol was cocked, and within a short distance of his head.

"Yes, T——, I take it up. I've nothing to do with your business matters, but while Soper is sick, in my cabin, no one shall lay hands on him if I can help it."

Retreating a few steps, resting the breech of his rifle on the floor, he said, "Mr. Steele I have nothing against you, —I've no quarrel with you."

"Nor have I any quarrel with you Mr. T——, but I do regret that you drew your rifle on me, and compelled me to get the drop on you; it might have been serious."

Nearly two weeks after, while working alone in a ravine, and stooping, some one fired at me, the ball entering near the center of the crown of my hat, and cutting off a lock of my hair, passed out at the side, lodging in a pile of soft sand and mud.

Rising up I saw smoke above a bunch of chaparral in the direction from which the bullet came, but could not distinguish any person. Whoever fired the shot was evidently concealed, and made his escape under cover of the brushwood.

Just over the ridge, on the main road, was Wallingford's saloon; upon inquiry I was informed that Mr. T—— while out hunting that afternoon had called there, perhaps to brace up his courage with drink.

I found the bullet; it was not bruised but unusually large for a rifle, and bore the marks of the grooves of the gun. Next an acquaintance borrowed the rifle which T—— carried that afternoon, and we found the ball was an exact fit.

Considering that this was the only rifle we could find with so large a bore, suspicion pointed very directly to Mr. T—— as the man who tried to take my life; possibly because of my friendship for Mr. Soper, or possibly he might have known that I had with me about five hundred dollars in gold dust.

But whatever the cause, I can testify that it is not pleasant to know there is an enemy with a rifle on your track. The more I studied the circumstances the more the awful fact appeared that their demand on Soper was premeditated robbery, and if it seemed to them necessary to accomplish their purpose, they would not hesitate at murder.

However, Soper's health improved, the winter passed quite pleasantly, and we did a fair business in mining.

During the winter an incident occurred that so illustrated the general character of the miners that I mention it.

In the fall measles were epidemic, and Mr. A. from Georgia came near dying with an attack; and when Soper had recovered so as to walk around, he found him in a neighboring cabin with greatly impaired health. Utterly dependent on his few acquaintances, so discouraged, homesick, and in despair of ever seeing his family, that there seemed no hope of his recovery.

We therefore concluded to bring his case before the miners, and see what could be done for his relief. Preparing a number of subscription papers representing his condition, and soliciting aid, we circulated them in the mining camps, and were immediately joined by others who appointed a time and place for the collectors to meet; and when they assembled, presenting their papers and gold, it was discovered that twelve hundred and sixty dollars had been raised.

The same evening Mr. A—— was informed of the action of the miners, and that there was nothing to hinder his going home just as soon as he was able.

He was quite overcome with emotion, and after expressing his gratitude, said, "I think I can start in the morning; I feel so much better, it's like coming back to life."

After a few days I went with him to Coloma, and informed the Stage Company that the miners were sending him home to save his life, and he was carried free to Sacramento, and, on the stage company's recommendation, received a free ticket to San Francisco, and in due time, with improved health joined his family.

## CHAPTER XIX.

*Indians as Miners.—Not even a Manger for the Babe.—Unwelcome Visitor.—Undesirable Bedfellow.—David and Jonathan.—William Hall Murdered.—Danger.—Sleep with Weapons at Hand.—California Cat.—Tchabo.—How his Name became Tom.—"All There."*

The Indians of the Columbia ranch were active miners in their way, using only a pan for washing, and, as they generally worked where the bed rock was bare, dug with their hunting knives in the slate. The squaws, always in companies of five or six, sometimes used a pick and spade, but I never saw an Indian use a rocker or long tom, unless at work with whites. Possibly they never had enough gold at one time to buy such machinery.

A party of five squaws worked below us in the ravine for several weeks; coming over from their village every morning about sunrise, and returning toward sunset. Two carried infants tied to a framework of boughs, and while the mothers were at work, the babies were hung on the swaying branch of a tree.

Near the place where they worked were a number of large bunches of chaparral. One day an old woman came and asked, in Spanish, for matches. Giving her some, she kindled a fire in the chaparral, and we noticed that some of them spent the greater part of the day there; but it elicited no particular attention, until they started for their village, when, as they passed us, we observed that they carried an extra baby. Not tied to a frame like the others, but in the arms, presumably, of its mother.

Evidently it had been born that day among the chaparral; and whether wrapped in swaddling clothes we knew not, but there was not even a convenient manger in which it might be laid. So with an endurance, which probably mothers only know, it was conveyed, more than three miles over the ridge to the village, in her sheltering arms.

Early in March, 1853, Mr. Soper discovered and opened a mine near the head of Kelsey canon, and as the winter rains were supposed to be about over, built a summer residence, covering it only with the leafy boughs of pine and live oak. After occupying it he was frequently annoyed at night with a rustling among the leaves over his bed. Believing it was made by an owl or some animal, and firing his pistol in the direction, it would immediately leave, but perhaps before morning the rustling of the leaves would be heard again.

It was easily frightened, but never seen or heard in the day time; nor could he find anything like a nest on the roof. Curious to know what it really was, he prepared combustibles, and when it was heard, quietly flashed them into a flame, and in the light was surprised and disgusted to see a large snake peering through the leaves just above his bed. Of course it was not such a companion as one would select for his bedroom.

Some time after, while spending a night with Soper, we heard a noise among the leaves, and a random shot brought his snakeship down. It was harmless, of the genus masticoplis, or coach whip, slender, and about six feet long.

Venomous snakes were seldom found in this region; rarely a rattlesnake, but frequently a colubrine, resembling the milksnake, and which would not hesitate to come into your camp or cabin, as the following incident illustrates.

After being away for several days, I returned to the cabin in company with Mr. Jeremiah Dobin. We had left our bunks with the blankets spread; and as it was about midnight, tired and sleepy, we undressed and lay down, when Dobin remarked "How does it happen that my blanket is wet?" and then, with a scream, sprang from the cot.

Lighting a candle, and examining his bed, we found that what he had taken for water, was the cold coil of a large snake (colubrine) against his leg, and when it began to wriggle for more room, of course he at once surrendered the entire bunk.

The harmless reptile was killed and cast out, and a thorough search satisfied us that no others were in the chinks, or about the cabin, before we could quietly yield ourselves to sleep.

Some time after Soper had gotten rid of his haunting snake, he said to me, "Don't think me superstitious; they say to dream of snakes indicates that you have enemies; but my snake was no dream; you know I have enemies; you heard one of them threaten to take my life. I never saw those people until we met in California. I thought they were friends as well as kindred, but have found them capable of any meanness. We may meet sometime, and I have to defend myself; it is my prayer that it may never happen; I shall avoid them if I can, but should I be compelled to shed blood, I might want your testimony in court. Will you please correspond with me so that I may know where to find you?"

I had only to refer to the bullet hole in my hat, and the many circumstances pointing to T——, to impress my mind with the thought that sometime I might need Soper's testimony as much as he could mine. It was another case of David and Jonathan. As a consequence we kept trace of each other for many years; even while serving in different departments of the Union army, and it was not until I became an itinerant minister among the Mexicans, that I lost his address.

I was now working alone, and washing gravel, some of which I had thrown up to disintegrate and dry, more than a year before. The water was failing in places, and many of the miners had gone to the larger streams.

However, I was doing well; too busy to be lonesome, until startled with the story that my neighbor, Mr. William Hall, had been robbed and murdered in his cabin, scarcely a mile from mine. I knew him well; a kind, steady, industrious, upright man. He crossed the plains from Missouri in the summer of 1850, coming into the mines by way of Placerville.

Last fall he sent his gold home by express, but probably had his winter's earning's in or about his cabin, and like myself was living alone. He was found by two men living at some distance, but working near him. Not seeing him at work, at noon they went to his cabin and found him cold in death.

Evidently he had been shot in the door of his cabin, the body dragged inside, and candle drippings indicated that the place had been searched in the night. We had no idea who committed the deed. Such things had been done near the towns, and in places visited by Joaquin's band, but this was the first in this part of the country, and I no longer felt safe to be alone in my cabin.

As they had no safe place of deposit, the miners had been accustomed to carry their gold with them. Usually in their coat pockets, taken with them to their work, and at night placed under their head. In buckskin sacks, men kept thousands of dollars, in gold dust, with them day and night.

Thus it came to pass, after a man had been a year or two in the mines, if he was industrious and temperate, it was supposed that he had gold; and if he ventured into certain localities alone was in danger of being murdered and robbed.

I now realized that an attack was liable any night, that no place was safe, and arranged my cabin so as to be sure that no one could enter without waking me; and then placing my pistol, butcher knife, and axe within convenient reach, all things being ready for defense, I lay down to sleep.

In one respect the danger did me good. It led me nearer to God. When reconciled through the atonement in Christ, placing all in His hands, feeling that whatever happened, God's love would not forsake me, sleep was sweet.

As my cabin was windowless I usually depended on the open door for light; and my table, against the side of the house, had a chink a little above which lighted it quite well.

One morning while at breakfast, a little California cat, brown, with dark stripes like a coon, crept in, caught a slice of bread from the table, and tried to escape through the chink, but the slice was too large, and in an instant I threw a towel over it, and, after a fight, made it prisoner.

Nailing slats across a box for a cage, I put it in, fed it some fresh beef which it readily ate. The next morning it ate from my hand, seeming quite tame; So nailing my blanket over the fireplace, and closing the chinks that it might not escape, I let it out of the box.

After exploring the room, it devoured a bit of meat, and seemed to want more, had no fear of being handled, but finally climbed on my shoulder, and began to pur, as though we were old friends.

Thinking it was domesticated, I opened the door and gave it liberty. It did not leave immediately, but came in, took another little slice of beef from my hand, and then started for the timber. Following it, I found where it lived in a hollow tree. It always remembered me, and coming often for food would linger around the cabin, but never permitted me to put my hand upon it again.

About this time *Tchubo*, the son of Capitan Juan, of the Columbia ranch, came and wanted to work for me. He was about sixteen years old, stout, and remarkably bright. I was glad to have him with me; felt safer, especially at night; and as he did fair work I paid him at the rate of five dollars a day.

Soon he was neatly dressed, as he said, "Como un Americano" (like an American.) Besides his native Indian, he had considerable knowledge of Spanish, and made wonderful progress in learning the English language. A smooth piece of slate about four feet long and three wide, placed on posts driven into the earthen floor of my cabin, had served as kitchen and dining room table, but now was used also as a kind of blackboard in teaching *Tchubo* to read and write.

He had learned that Juan, his father's name in Spanish, was the same as mine, John, in English, and so he wanted me to tell him what his name was in English. This puzzled me, but I finally said "Thomas, or it might be only Tom."

This pleased him, and he repeated, "Juan, John, Tom." And insisted on being called by the English name, Tom.

When not at work, he usually employed his time in making letters, and pronouncing them in English, and by the constant repetition of words, phrases, numbers and sentences, he became quite proficient in the English language.

His father often visited us and seemed much pleased that his son could talk English, and was learning to read and write. I became greatly interested in Tom. His older brother had been killed in battle with the Placerville Indians, and I thought when he took his father's place as chief, he would introduce educational methods, and perhaps through Christianity save his people.

Another thing, they were not beggars, but did considerable mining, and expected to pay for what they received. One day leaving Tom at work, I went to the cabin to prepare dinner, and was surprised to find several Indians there. They had found my tin box containing about fifty dollars in gold dust, and had poured it on the table to look at it.

When they saw that I wanted to use the table, one swept the gold into the box, saying as he did so, "*Todo hay*," all there). Curious to know whether any had been taken, for it was a great temptation, when they were gone I weighed it, and found it, sure enough, "all there."

To me these Indians were not only friendly, but, as far as I knew, truthful and honest, and I felt perfectly safe in person and property to the extent of their ability to protect me.

## CHAPTER XX.

*Beginning of Trouble.—"Never Anymore." Not Brandy, Whiskey.—Money All Gone. —Murder.—Tom in the Hands of Savages. —The Funeral.—"Somebody Must Die." —Trial and Execution.—Capitan Juan's Speech —Tom's Philosophy.—"You'll be Delivered Up to Your God."—Tom's Ruin. —Sad Fate of Many.*

Tom usually spent Sunday at home, returning either in the afternoon or early Monday morning. If for any reason he wanted to remain away a few days he always let me know in advance; and so when he failed to appear one Monday morning I felt somewhat anxious; but towards noon he came from the direction of Wallingford's saloon, his unsteady step revealing the fact that he was drunk.

He wanted to dig, but was utterly unable, and at last I coaxed him to go with me to the cabin, where he lay down and slept until near supper time. Awaking and coming where I was at work, he complained,

"*Tom muy enfermo*" (very sick) "sick, heap sick."

"Yes," said I, Tom you've been drunk, *muy borracho*, (very drunk) what did you drink?"

"*Aguardiente.*" (brandy.)

"Who gave it to you?"

"Wallingford."

I remarked, "It will make any body sick, make them do what they don't want to, and make them so they don't know what they do. You Tom, never drink any more *aguardiente.*"

"Never any more, never!" said he with emphasis; placing his hand upon his stomach, and soon turning away, vomited freely. Coming to the cabin, and sipping a little coffee, he was finally able to take some supper.

He said he had never tasted *aguardiente* before, and did not know it would make him sick. In talking with his father, as well as from his own statement, I do not think Tom ever before tasted spirituous liquors.

One morning, a week after this, I sent him with a note, as I had often done, to Mr. Cooledge, cf Peru, for groceries. It was afternoon when he returned, drunk, and without the groceries. When questioned, he said they were at Wallingford's.

As soon as he became quiet, I visited Wallingford's; they told me that he had stopped there on his way from Peru; engaged in gambling, drinking freely of whiskey, and finally left without taking his package, had probably forgotten it. Explaining the matter, I begged of them as a favor to me, as well as to him, not to let him have liquor of any kind.

The next morning I questioned him as to drinking *aguardiente* again, and he promptly replied, with perfect candor,

"*No, no he bebido aguardiente, bebi whiskey.*" (no, I did not drink brandy, I drank whiskey.) This accorded with Wallingford's statement, and I could not resist the impression that Tom had been deceived in the name, and doubtless would have refused brandy.

It reminded me of the proverb: "Wine is a mocker, strong drink is raging, and whosoever is deceived thereby is not wise." And I was aware that many young men, with greater advantages than Tom had been deceived thereby. I tried to explain that whiskey made him drunk the same as brandy, and that rum, gin, ale and wine were just as bad. Still, I entertained the hope that Wallingford was manly enough to refuse him liquor.

Next questioning him about his money, taking the empty purse from his pocket, he replied, "All gone."

"Who got your money?"

"*No se,*" (I don't know) one man at Wallingford's." Of course, that was all they wanted of Tom, and I suppose he gambled it away.

The next Saturday afternoon Capitan Juan made us a visit. After supper I paid Tom his earnings, $15.00, and saying he would return Monday morning they started for home. Knowing that Capitan Juan never drank, I was glad they were together, feeling that his son was safe with him.

Late Sunday afternoon Tom returned; there was blood on his clothes and hands, he was greatly excited, but too drunk to tell just what had happened. Flourishing his hunting knife he tried to show me how somebody had been killed; but who it was, or who did it, I could not make out, but began to regret having anything to do with him.

At last he fell into a deep slumber, and did not wake until morning. Then questioning about the blood, he seemed to remember all that had occurred from the time he and his father left my cabin until his return.

On the way to Peru for provisions, passing Wallingford's saloon, he came out and invited them in; Capitan Juan refused to enter, and then Wallingford gave Tom a small bottle of some kind of liquor. His father took it, smelled of it, and at once dashed it in pieces against a rock.

The next day Tom, and two other Indians, came back to the saloon, engaged in gambling, and all drank freely of what Tom called "*vino*" (wine); and when all their gold was gone, started for the village. The two Indians quarreled, one stabbing the other with a knife.

Tom tried to help the wounded Indian home, but he died on the way. And then, Tom, afraid of his father's anger, should he appear before him drunk, came to me.

Afterward I learned that the one who committed the deed, told of it in the village, and the friends went out, brought in the body, and prepared for the funeral.

Now the fact dawned upon my mind that in paying Tom wages, and teaching him English, instead of helping him, as I had fondly hoped, I had put him into the hands of the worst kind of savages.

While he had no money he was comparatively safe from the saloon keeper, but when they saw him well dressed, and he understood enough English to tell them in answer to their questions, that he had earned the money, paid for his clothes, and had money left, all the wisdom "of that old serpent the devil," was invoked to compass his ruin. Not that they had anything against him; they simply wanted his money, and gladly descended to the lowest depths of merciless meanness in order to get it.

With Tom I attended the funeral of the murdered man. Entering the village we passed a man sitting alone at the door of a wigwam.

"That's the man," said Tom. "He killed him, and must die."

"Why don't he go away?" I inquired.

"Because if they not find him, they take his father, his brother, his son; somebody die—you see."

Afterward, upon inquiry, I found that this was a kind of common law among these Indians. When murder was committed, the murderer must be put to death. If he ran away, the nearest of his male kindred, whom they could find, must die in his stead within a year. But Capitan Juan said he never knew a case where an innocent person was executed for his kindred's crime.

The funeral was inexpressibly sad; every countenance and voice indicated genuine sorrow. The body was burned with the usual ceremonies; and after they were concluded, twelve warriors with bows and arrows slung over the left shoulder, and each with a long flint-pointed spear in the right hand, in charge of two officers, came suddenly upon the scene.

Marching with them, unbound, was the man who had killed his companion. Passing to the front of the chief's wigwam, they halted, and their prisoner was seated on the ground. Capitan Juan came out and after talking awhile in Indian, also sat down.

Then followed something like a trial, conducted by one of the officers. Tom was called and questioned; others came forward and spoke. It seemed as though the entire village, with a number of whites were present. The guards kept a large space open, so that the chief, officers, prisoner and witnesses could be seen.

Again the chief arose, and with great earnestness and solemnity spoke at length. The Indians were deeply affected by his talk; and how I did regret that I did not understand it.

When the speech was ended, two men approached with cords of some kind of bark, bound the prisoner's hands together at the wrists; another cord around the elbows was tied across the back, and the limbs were bound firmly together at the ankles and knees. He made no resistance, and I am not aware that he spoke during the operation.

Then a blanket was thrown over his head, a cord placed below his knees, and over the back of his neck, drawing his head and knees together, and so, with the blanket bound closely around him, he was carried outside the village, where a funeral pile of dry wood had been built, and already fired, and he was laid thereon.

Sick at heart I turned away from the awful scene; realizing, as I did not before, why the deep sadness everywhere seen and heard. But also convinced that the untutored Indian, who, in the delirium of intoxication, slew his friend and met his cruel fate, was less guilty than the "civilized" white man, who, knowing the probable conseqnences, yet to obtain the paltry half ounce of gold, furnished the liquor and tempted him to drink.

By questioning Tom, I obtained an idea of his father's address. He reminded the Indians that these two men were good men, brave warriors; they had families who loved them, everybody respected them. They were friends; friends always like brothers; would always have been so, only for the white man's drink. It makes people bad. If white men drink it, it makes them bad; if you drink it it will make you bad; if I, (Juan) drink it, it will make me bad. It is bad, always bad, very bad.

"Tom," said I, "do you think your father is right?"

"Yes," said he, "white man's drink always bad; it makes me bad when I have it and don't drink it."

"How is that, Tom?"

"When my father took the bottle and broke it, I was *muy enojado* (very angry) at my father, and come back next day to get more; I ask them to come; now both dead; Tom bad."

Doubtless Tom's philosophy was correct, and there was encouragement in his candor and self-accusation; and I fondly hoped he might become as bitterly opposed to the use of strong drink as his father. But alas for Tom; he had already acquired that unquenchable thirst; and there were those who, in its most seductive form, put the temptation in his way.

It was not long until Tom was drunk again. Determined if possible to save him, and knowing it was contrary to law to give an Indian liquor, and that the Indian affairs were in the hands of the War department, I addressed a letter to General Hotchkiss, then in command of the Pacific Department of the army. Promptly receiving a reply, with necessary instructions and blanks for making out a complaint against any one selling or giving intoxicating liquors to Indians.

But now unexpected difficulties arose; no Indian's testimony could be accepted; and no white man, who knew the facts, was willing to testify.

Again I appealed to Wallingford, urging that to whomsoever he sold liquor, let none be given to Tom. He had probably been informed of my effort to invoke the law, and replied with a volley of profane and insulting epithets, and ended by saying, "I understand you are trying to make trouble; better let it drop. You attend to your little business, and we'll attend to ours; and, mark my words, if you interfere with our affairs, you'll be delivered up to your God."

After this Tom became worse; whatever he earned went for liquor. Losing his former self-respect, he worked but little, and would beg for money to buy liquor; and, to the grief of his father, spent his time among the saloons.

Other members of the tribe became equally debauched, until it seemed as though it would have been far better had they all perished with smallpox. And I believe that the sad fate of Columbia Ranch overtook every Indian village in the California mines.

## CHAPTER XXI.

*At Coloma Post Office.—The Preacher.—Under the Great Live Oak.—Card Playing.—Antoine Canon.—Snow Storm.—Prof. Hamilton.—The Devoted Wife.—Terrible Tragedy.—The Orphan Boy.—Robert Neale's Charge.—Finds a Home for the Orphan.*

Sunday, May 1st, 1853. Yesterday afternoon Tom went home, and I visited the post office at Coloma. Mail from the last steamer had just been distributed, and as the news spread, as usual, people flocked in from all sides.

Some were made glad with good news from home; others anxious, because the expected letters had not come, usually tried, after the style of an auctioneer, to buy a paper containing the general news from their part of the country; and it was common at such times to hear the exclamation, "Who has a paper for sale from New York?" or from such and such places; and people receiving papers, after their perusal, sometimes sold them for fifty cents or one dollar.

Some receiving bad news, went sorrowfully aside to weep.

A little below the post office, in the shade of some pines, people often retired to read their letters and papers. Paying a dollar for a copy of the New York Weekly Tribune, I went thither and sat down to read.

Soon a fine looking man stood up, and with a voice of wonderful power and compass, sang the hymn beginning,

"O for a closer walk with God."

People were attracted; gathering as silent, attentive listeners. Then, by invitation, a number joined with him in singing,

"Jesus lover of my soul."

The effect was inspiring; and there kneeling down he offered a prayer which made me feel that God was not only present, but considering our individual interests. Next, drawing a Bible from his pocket, he read the fifty-fifth chapter of Isaiah, with a few verses from the third chapter of John; and preached from the words of Isaiah, "Seek ye the Lord while he may be found, call ye upon him while he is near."

I was very much impressed with the sermon, and glad to learn that he would preach at ten the next forenoon near Downing's ravine.

Some interested men took sluice lumber, as yet unused, and prepared very comfortable seats in the shade of a great live oak, and at the appointed time a large congregation had assembled. Of course there were no women or children.

Several old hymns were sung reminding us of home, its associations and worship; several earnest prayers were offered, and then the minister preached a clear, earnest and impressive sermon from the one hundred and sixteenth Psalm. Many were moved to tears; and in closing he announced an appointment to preach at five that afternoon in Coloma, but if they would remain, he would hold another service immediately after dinner in this place. All seemed desirous for this, and as my cabin was near, he accepted an invitation to dinner.

We were gone about an hour and a half, and returning, came over a low ridge among clumps of chaparral, which concealed us until we were within a few rods of the seats, when, looking down upon them, a scene utterly unexpected met our view. Not all, but a large part of the congregation, divided into groups of twos, fours, perhaps eights, the members of each group facing each other, and using the seats for tables, were busily engaged in playing cards.

I don't know whether there was any money at stake, but just as soon as the minister was seen to be present, the cards were quietly pocketed, and all assumed the attitude of serious attention.

Never, before or since, have I seen or heard of a congregation, while on the Sabbath waiting for the minister, engage in such a pastime, and I have always believed, had there been women and children present, the exercises would have taken a more intelligent and spiritual trend.

After another earnest discourse on John twenty-first chapter, and twenty-second verse, the words of Jesus to Peter, "What is that to thee? follow thou me;" and a fervent prayer, he went his way. Perhaps he was disappointed; may have felt that his words were lost in the echoless air, but I am sure many were helped by being led back to the home life and associations, by seeing the true life in Christ, and made to realize personal responsibility, by being brought face to face with God.

If I learned that minister's name, unfortunately, it was not written in my journal, and has been forgotten, but have thought he might have been William Taylor, afterward the world wide missionary, and Bishop of Africa.

Monday, May 9th, 1853. As the dry season advanced and water began to fail on the uplands, closing my mining operations at Downing's ravine, with James Badgely, John Berry, Levi and George Chapman, brothers, and George Ward, all recently from the Georgia gold mines, I started for Antoine canon, far into the mountains between the north and middle forks of the American river.

With provisions, blankets, and mining tools, we also carried a tent, eight by ten feet in size, and lest the snow might be too deep on the mountains for mules, also saving expense, we went on foot. It was our intention to make a reconnoissance, as a soldier would say, and if circumstances were promising, occupy the place and devote the summer to mining.

The snow was deep and quite hard on the ridges, but descending into the canon on the evening of May 11th, found that it had mostly melted away. We pitched our tent, and the next morning made an exploration of the place. A number of houses had been built here last summer, all being vacated in the fall, and every roof had been broken in with the weight of the winter's snow. The water was too deep for successful mining.

About noon a heavy rain set in, and we took shelter in the tent, which was very comfortable, until toward night when the rain turned to snow, which accumulated so fast that the tent was soon in danger of collapse. At intervals by shoveling off the snow, we relieved it of the heavy strain; but the snow continued all night, and most of the next day, when the air became decidedly colder.

Snow in the canon had fallen to the depth of about four feet, and, of course, we could not hope to begin mining within a month; so wading, or rather wollowing out, we came down to the middle fork of the American river. On our way we stopped awhile at a mining camp known as Yankee Jim's. Here I found Prof. Hamilton, of an eastern college, with whom I first became acquainted in Onion valley, and he related the following terribly tragic incident, which occurred a few weeks before.

The reader will remember Doctor Y. mentioned in chapter IX, among those who were snow bound in Onion valley in the winter of 1850 and 51. Graduated at West Point, he held the position of surgeon in the United States army; served through the Mexican war, but owing to his intemperate habits, after the battles around the city of Mexico, was returned to New Orleans and discharged for drunkenness while on duty.

Ashamed to go to his wife, who lived in Kentucky, he ceased correspondence with her, and drifted into California. In some way she learned where he had gone, and why he lost his position in the army. Knowing his proud spirit, she suspected the reason of his silence and absence, and with woman's love, constancy and devotion, resolved to save him if possible.

After addressing letters to him, telling of her intention, she went to Sacramento, where he joined her, established their home and practiced his profession over a year. But falling into his old habits, doubtless to get him away from his associates, she persuaded him to go out into the mountains; and so they came to Yankee Jim's, where, being a skilled physician and surgeon, and the only one in that region, he entered upon a lucrative practice.

For some time they were prosperous and happy, but again his old enemy overtook him. His wife, an excellent Christian woman, with their little son, about a year old, appealed to his better nature, and for a time sustained his nobler manhood in the desperate struggle to assert itself. But while he had money, men, who knew his weakness, plied their temptations beyond his power to resist.

Gambling was added to drunkenness, his earnings were soon gone and he was reduced to want. His wife was neglected, sometimes abused. God only knew the burden of that devoted heart; away from congenial society, all her efforts vain, every cherished hope dying out; but she never gave him up, nor faltered in her efforts to save him.

One night, becoming troublesome, he was ejected from a saloon. Maddened, and in a fit of delirium, he came home, took his gun and threatened to shoot his wife. By accident he overturned the candle and put it out: in an effort to escape she ran to the door; opening it the moonlight revealed her form, and he fired, killing her instantly.

The miners, having great respect for her, incensed at his awful deed, at once hung him to a tree, and probably not until the tragedy was complete did they realize the presence of the orphan child. The little boy, scarcely a year old, and afraid of everybody except Robert Neale, a boy about sixteen years of age, who, employed by the Doctor, had come from Sacramento with him and his wife, and was present in the house when she was killed.

Poor Robert, left with such a responsibility on his hands, and not a woman anywhere in that section of country to whom he could go for advice or help, was quite overwhelmed.

Fortunately the child was greatly attached to him, and he knew how to prepare some kinds of food for it; but of course it missed its mother, and again and again cried itself to sleep.

Prof. Hamilton kindly assisted Robert and the baby all he could, remaining with them in the house. He said the next night after the mother's death, the baby was restless, cried frequently, and Robert carried it in his arms nearly all night. At daylight the place became quiet, his step was no longer heard, so the Prof. peered into the room; there they lay in sound slumber, Robert holding the baby tenderly in his arms, and both faces showing traces of tears, as though they had cried themselves asleep.

Robert proposed taking the child to a family living near Sacramento, acquaintances, perhaps relatives of his. The miners helped him to arrange for the journey; so taking the Doctor's horse, and safely carrying his precious charge, he made the trip to the valley, and there in the family it found a home, protection and care, never realized any where else.

But Robert Neal had become so attached to the little orphan, that he concluded to remain near him, and wrote Prof. Hamilton that he had found employment on a neighboring ranch.

## CHAPTER XXII.

*Return to Downing's Ravine.—Visit from Capitan Juan and Son.—Invited to an Indian Council.—Eight Villages Assemble.—Place of Meeting.—Boiling Water in Baskets.—Why Pah Ute Indians were Called Diggers.—Indian Dance or Drill.—Interesting Game.—The Council.—Settling Difficulties between Villages.—Liquor Problem.—Inside View of Indian Life.—Political, Social, Religious.*

Returning to Downing's ravine I received a very cordial visit from Capitan

Juan and his son Tom, who informed me that representatives from eight Pah Ute villages would soon meet with theirs in council, and invited me to be present.

Of course, like most boys, I was more or less ambitious; but to succeed to the dignity of a seat in a council of Indian chiefs, was something which neither my age nor ambition, up to that time, had ever suggested. However, I made a special effort to be present.

The place selected, about two miles north west of the Columbia village, was prepared by placing a row of small poles firmly in the ground, inclosing a circle about six rods in diameter. Into these were closely woven rods of chaparral to the hight of over eight feet.

In the center was built the council fire, around which the chiefs smoked and deliberated.

Two entrances, one east, the other west, were so arranged by the braided walls over-lapping, with a space of about three feet between, that at a short distance the passage was not visible, and a person might walk entirely around the corral and see no opening.

Near the west entrance were fires, and a number of squaws engaged in cooking for the feast. They had cone-shaped baskets made of split wood fiber woven very firmly together, gummed with some kind of resinous substance, so that they not only held water, but water could be boiled in them.

This was effected by placing the baskets in pairs, the points in the ground so as to keep them steady. Water was poured into each; if they wanted to boil meat, it was put into one, and a hot stone, taken from the fire with a green branch bent like a pair of tongs, and nicely rinsed in the other, dropped in with the meat. When it cooled, it was taken out and another hot one put in, and in this way the water was kept boiling until the meat was cooked.

Usually, one basket of water sufficed as a rinsing place for the hot stones which kept many others boiling. So also they cooked a kind of paste, made of flour and water.

I noticed, however, that many brought their provisions with them. Some had large cakes made of acorns ground quite fine, after removing the shells, then mixed with grasshoppers and baked on a hot stone. Another article of diet, wild clover (alfafa) ground very fine while green and baked.

They also use a brittle vine, in taste resembling lettuce; but especially in the spring, their principal food was a bulb of pleasant taste, about the size of a plum, growing just below the surface of the ground, the place indicated by its fringe-like stem, only a few inches in hight.

Often bands of Indians were seen traversing the slopes, each with a pointed stick, digging and eating these bulbs. Doubtless from this practice they obtained the name "*Diggers*."

On the eastern side a large space was smoothed off, and used for their drill or war dance, which continued day and night for eight days.

I was told that there were about eight hundred warriors present, though I think only about one hundred drilled or danced at the same time.

They began by forming concentric circles around the leader, each one holding a spear or long stick in his right hand; facing inward they followed his motions, raising their spears perpendicularly, and bringing the handle with a thump on tne ground, meanwhile singing in unison, "*Hah, hi-yah; hah, hi-yah; hah, hi-yah;*" and at a sign from the leader they would all face outward, still keeping up the motion and the song. Then facing right or left they would march in circles; and at another signal the circles would be formed into squares, and, all facing in the same direction, would march to another place near by. When one set became tired, another took its place, and thus the performance was kept up.

A rather interesting game was played by two parties, each numbering about twenty. An open valley was selected, and two trees generally about forty rods apart, were designated as belonging one to either side. The persons in the play, each with a stick about four feet long, met midway between the trees. A belt, about four inches wide and three feet long, made with strips of buckskin, cloth or bark neatly braided, was thrown high into the air.

After this no one must touch it with his hand until it had come in contact with one of the designated trees. It was tossed on the sticks, and the party bringing it first to their tree were winners.

It was a very exciting game and occasionally the belt would be tossed several times around both trees before touching either. Of course, in the wild rush some would be hurt, but I never saw any indications of anger among them. On the contrary they would laugh and shout, and though suffering intense pain, were generally successful in avoiding any outward evidence of it.

At this meeting the leading members of the various villages became acquainted, difficulties and misunderstandings were defined, brought before the council, and if possible, settled to the satisfaction of all. In this manner peace was concluded between the Placerville and Columbia villages, and when finally they separated there seemed to be universal friendship and good will.

Among other things they discussed the liquor problem, and were greatly sur prised when I told them that the government made laws to protect them, and would punish any one known to give or sell liquor to them; something it did not do even for its own people.

It is a fact which needs to be emphasized that the government of the United States enacted laws for the protection of the Indians which were just and humane, and it has been in direct violation of these laws that the Indians were defrauded, debauched and often murdered. And those who engaged in this work of outrage, were not always citizens of the United States; and though sometimes Christian in name were not so in fact, but were too low, morally, to hold membership in any protestant church.

Under our constitution and laws, people enjoy greater liberty than in any other country. Like the mercy of God, who sendeth the rain and sunshine upon the just and unjust, so personal liberty is given whether the recipient is worthy or not. And it is these who pervert the liberty of which they are not worthy, that have brought destruction upon the Indians, and disgrace upon the government and people of the United States.

Odium still clings to New England because many years ago witches were hung there; and yet no witch was ever hung in accordance with any law made in America; and so it may be said, no Indian was ever corrupted according to any law of the United States. Our disgrace in this matter has been through the recognition of foreign customs and laws, and a criminal disregard of our own.

Most of the miners had gone to the low-lands and rivers, and I was afraid that when so many Indians assembled, they might obtain liquor, become intoxicated, and not only fight among themselves, but provoke a war of extermination by an attack upon the whites. However, during the entire eight days I did not see an intoxicated Indian.

Perhaps those who sold liquor, had a wholesome dread of drunken Indians; and the Indians themselves exercised a restraining influence upon each other. While there were doubtless many, who, had they been tempted, would have drunk to excess, yet the common sentiment was against the use of strong drink. And it is to be regretted that there are so many cities and villages in the United States where to-day the temperance sentiment is lower than among those Indians at that time. Even in the council, were chiefs whose bloated faces and bloodshot eyes indicated that they were victims of drunkenness, but without a dissenting voice the white man's drink was condemned. Not even appetite, profit, companionship, party or prejudice had taught them to prevaricate, but with perfect candor they innocently told the truth.

While the deliberations of the council were in the Indian tongue, they all possessed some knowledge of Spanish, and took special pains to have all matters interpreted for my benefit. And conversing freely with Tom, who, as the chief's son, and prospective heir, was present at all the sessions, I obtained an inside view of their political, social and religious life.

The villages were generally organized by the election, common recognition, or selection by the chief of four officers.

1st. Captain of the Warriors; whose duty was to organize and drill the men as soldiers; and, under direction of the chief, lead them in time of war.

2nd. Captain of the Boys; teaching them to make bows and arrows, to hunt, fish, endure fatigue, practice proper self-restraint and etiquette. The boys were generally under severe discipline until old enough, or rather, big enough to be recognized as warriors.

3rd. Superintendent of Works; having charge of industries, such as gardening, collecting and distributing supplies, and involved the temporal prosperity of the village.

4th. Master of Ceremonies. A kind of priest and civil judge, before whom marriages were recognized; difficulties brought for adjustment, except high crimes or appeals, which came before the chief; and by whom funeral ceremonies were conducted.

These four, with the chief, formed a council of state, and were in fact the government. All their services were rendered gratuitously, and as for gaining a livelihood they seemed to have no advantage over the rest.

I noticed that difficulties between villages had been brought about, not by any general difference of opinion, as to boundaries &c., but by individual misdoing, shielded by the personal friends of the culprit; which acts, when pointed out, were recognized as wrong, but in no case was there any demand for the punishment of the criminal.

It appeared that the recent war between the Placerville and Columbia villages originated in a quarrel between two boys, one from each village, both equally wrong, but supported by their friends.

In social life polygamy, though unpopular, was allowed, but it did not appear that any member of the council had more than one wife. There seemed to be a decided sentiment that men and women must be morally above reproach. And perhaps it was because of this that the squaws were associated in bands, and rarely, if ever, seen alone.

All were devoutly religious; or perhaps it were better to say superstitious. Some could recite, in Spanish, parts of the litany of the Roman Catholic Church, were decorated with crosses, carried beads and *resarios*; but their ideas of God were not different from others who were veritable heathen. They imagined that each village had its individual god, or at least, the Indians and whites had different gods; and while possessing wonderful power, the Indians evidently were not sure that any of them were good. They might be propitiated, but might also be very unreasonable and cruel.

They certainly feared their gods. Surrounded by superstitious dread of the unreal, they were ignorant of the real.

Trembling before imaginary gods; ignorant of the God of love, their sad lives were made sadder by contact with corrupt men, who by their vices led the masses to terrible and hopeless ruin. Yet, in the Indian mind, these men were associated with the name and religion of

Christ, which greatly perplexed the more intelligent and pure natives.

## CHAPTER XXIII.

*Desire to Renew my Studies.—Letters from Wisconsin.—Prepare to Leave California.—Last Interview with Capitan Juan.—Goodbye Tom.—Visit from P. O. Soper.—Meet with Uncle B. H. Robinson.—Reach San Francisco.—"Eternal Vigilance the Price of Liberty."—Shadowed.—Trapped.—Escape.—Corrupt City Government.—Murder of James King.—Reign of Terror.*

Saturday, June 18, 1853. I visited the post office at Coloma. Many of my most intimate acquaintances had either left California or I had lost their address, and a lonely feeling came over me; and a desire to renew my studies, turned my thoughts toward home.

Letters from Wisconsin, especially one from brother Edward D., a student at the University of Wisconsin, stating that during vacation he expected to visit our old home in New York, and urging me, if possible, to meet him there, confirmed my desire to return at once.

In a few days my affairs were arranged for the homeward journey. Meeting with Capitan Juan I told him that I was going to New York.

"*Cuando quiere volver?*" (When will you return?)

"*No se, quisa nunca.*" (Don't know; perhaps never.)

"*Donde esta Tchubo, su hijo de V?*" (Where is Tom, your son?)

He replied; "*Quien sabe?*" (Who knows?) Then he told me that Tom had been drinking, and was having trouble.

I expressed regret, hoped he might reform, and closed by saying, "*Vuestro merced*" (your excellency) a title of respect with which I always addressed him, knows that I would be glad to help him reform if I could."

"Yes," he replied, "I respect you, every one in the village respects you; you have been a brother to *Tchubo*, (Tom) and he loves you as his best friend and brother; but," he added with deep emotion, "Nobody can help him,—*Mi pobre hijo perdido! me pobre hijo perdido!*" (My poor lost son! my poor lost son!)

Our parting was very sad, and as I bade him *Adios*, he gave me an affectionate embrace, and placing his hands upon my head solemnly said, "*Dios te guarde por todos los cominos de V.*" (God keep thee in all your ways.)

I was much impressed, and wanted to find Tom; at our last interview he said he did not intend to drink any more, and I wanted to have one more talk with him on the subject, so that when he thought of me, he would always remember my desire for his reform.

While crossing the plains in 1850, at the camp of the mountaineer, Jim Baker, I bought a suit of Indian made buckskin. The frock coat, neatly belted and heavily fringed, I had kept as a souvenir, but just before the council, Tom's clothes being somewhat shabby, I gave it to him. It was a good fit, and being of Indian work was more appropriate than any other style of dress. He was greatly pleased with it, and he having pleasant features and fine form, it was much admired.

In my search for Tom I saw several bands of Indians, but not seeing his coat passed them at a distance. At last near Peru, meeting an Indian I inquired in Spanish, "Have you seen the Chief's son?" He replied, pointing in the direction, "Yes, he is down there by the road."

Going to the place, I found him reclining against a fallen tree, in a drunken stupor. His buckskin coat and flannel shirt were gone, even his pantaloons had been exchanged for a torn and ragged pair. Evidently "he had fallen among thieves," who had stripped him of his raiment, &c.

As I looked into his bloated, besotted face, and remembered what he had been, his bright mind, noble ambition, studious habits and untiring energy, before the demon alcohol had done its work; his despairing father's words seemed to ring again in my ears "*Mi pobre hijo perdido*," (My poor lost son.) Tears filled my eyes as I lingered beside him, deplored his ruin, and found it inexpressibly hard to give him up.

Does any one say, He was only an Indian; yet there were infinite possibilities bound up in his life. And although thousands in all ages have met a similar fate, still sorrow over one is none the less bitter on that account.

Without trying to awaken him, with a silent good bye I turned away and never saw him again. What became of him I know not; most probably he died a drunkard. However that may be, doubtless in the day of judgment it will be more tolerable for him than for those who lured him to ruin.

Greatly depressed I returned to my cabin, and about dark was cheered by the presence of Mr. Soper, who spent the night with me, and together, the next morning, we went to Coloma.

Here I learned that an uncle, Mr. B. H. Robinson, of Prattsville, New York, who years ago had spent some time in California, had just returned, and was at Uniontown, two miles down the river. For awhile I was inclined to change my plan, and remain in California, but as my business interests were arranged, concluded to go on. So bidding good bye to Soper, who returned to my cabin; and after spending an evening with my uncle, whose family I expected to visit in New York, a stage ride of sixty-two miles brought me to Sacramento, where taking a steamboat, sometime during the following night reached San Francisco.

I had planned to sail about the last of June, Monday, the 27th, on the steamer Sierra Nevada, and was detained but a single day in the city; but in the early part of that day an incident was added to my experience, which emphasized the adage, "Eternal vigilance is the price of liberty," and other valuable things.

On leaving Uniontown a burly, rather well dressed man occupied a seat in the stage; we took dinner at the same hotel, and when I purchased a ticket on the steamer for San Francisco, he also was present and obtained one. Though a total stranger, he knew that I had been for some time in the mines, and that I was now on my way home.

Surprised at his knowledge, for I had told him nothing about it, but supposed he might have overheard the conversation with my uncle the night before, and therefore it excited no suspicion. As he was social and pleasant, professed to be well acquainted in San Francisco, and suggested a stopping place; but I had made up my mind to stop at the Atlantic hotel, as I had been there before. "Yes," said he, "that's a good place, and I'll go with you."

However, there were several Mexicans on the boat, and wishing to improve my Spanish, I spent most of the time during the trip in conversation with them.

At midnight, or later, our boat reached the wharf, and taking our satchels we started for the hotel. When almost there he invited me up to a lighted room, "to have something to drink."

"No, I wanted no liquor."

"Then will you be so kind as to wait here until I return?"

I begged to be excused, could be of no special service to him, and was quite ready for sleep. I was surprised in not finding him next morning at the breakfast table; in fact he had not come to the hotel.

Looking over the morning paper, I noticed that the Sierra Nevada lying at long wharf, was announced to sail the next day; so, preparatory to the purchase of a ticket, I started for the steamer to select a state room.

It was early in the forenoon, and on a thronged street, when I again met the stranger who had accompanied me from

Uniontown. We recognized each other, and in passing he unawares crowded me against a door, which was on a level with the side walk, and, with a sudden push, thrust me inside.

Instead of the usual revolver, I carried two single shooters in a place prepared inside my coat; and, while with my right hand trying to prevent his shutting the outside door, with my left hand I cocked one, drew it, but just then saw another man standing in a side door, and as I raised the pistol he disappeared and shut the door. In an instant I drew the other pistol with my right hand, when the man who had pushed me in disappeared through a door on the opposite side and it was shut.

Bewildered, I stood for a moment with a cocked pistol in either hand, and on regaining presence of mind, saw that the room was only about six feet square, but containing three doors. Coming in from the street there was a door on the right and left, through which the men had disappeared.

Approaching the front door, which my assailant, in his haste to get beyond the range of my pistol, had failed to close tightly I swung it open, and stepped out upon the side walk.

Meeting a policeman I asked him to arrest the man who had assaulted me.

"Where is he?" he asked.

"In this house," I replied.

"You can't identify him."

"Yes, he followed me all the way from Uniontown, I can't be mistaken in the man who laid hands on me."

The policeman paid no further attention to my request; so congratulating myself that I was still alive, in possession of my liberty, passage and expense money, went to long wharf boarded the steamer, selected a state room and going to the shipping office secured a ticket.

As I reflected on the episode of the morning, the fact that I had been pursued by a robber became apparent, and only instant resort to the pistols saved me from being robbed or worse. The room, into which I was so suddenly pushed, was evidently a prepared trap, into which the victims who could not be decoyed might be forced.

But even with this experience, I had no idea of the actual condition of the city. The city government at that time was entirely in the hands of the saloon element, gamblers and thugs.

Up to this time more than twelve hundred murders had been known and registered, and there were reasons to believe twice that number had been committed; and yet not a criminal had been brought to justice.

Policemen, police courts, officers of all grade were implicated in crime, even to Judge Terry of the United States district court.

Men absorbed in their business affairs had neglected their duties as citizens, and the baser sort had taken possession of the offices, and those whose duty required them to protect the people, and the legitimate business of the city, became a terror and menace to both.

There may have been some well-meaning men in office, but they were too few to exert any influence, and at this time San Francisco was governed by criminals, and the people lived in fear.

A few months later, the editor of the Evening Bulletin, James King published an article which displeased a prominent official, and was deliberately shot down, at noonday in a busy thoroughfare, the murderer making no effort at concealment; so confident was he that no court in the city would convict him.

There was a veritable reign of terror, with life and property at stake, men were afraid to offend the officials, and at the same time dared not trust each other.

## CHAPTER XXIV.

*The People Organize for Defense,—Vigilance Committee.—Officials of San Francisco Captured.—City Fortified.—"Government of the People, for the People, by the People."—Restoration of Order.—Plan a Journey through Mexico.—Gen. Santa Anna Interferes.—Sail from San Francisco.—Fourth of July at Sea—An Anxious Night.—Acapulco.—Change Plan of Journey.—Carried Ashore at San Juan del Sur, Central America.—Honesty.—The Original Currency.—Virgin Bay.—Living Gems.*

At this time many citizens of San Francisco regarded its government as hopeless; not only the offices but the ballot boxes being in unworthy hands. To avoid collision with the officials, and endure what they seemed powerless to remedy, was the best they could expect.

But there was another class who attended strictly to their own business, and were desirous that others should do the same. Patiently enduring evils, while they were endurable; but if compelled to suspend their business operations in order to chastise evil doers, they were not embarassed as to methods; doing it quietly, effectively, and with the least expenditure of time.

Possibly at this date, in proportion to its population, California had a greater number of this class than any other state.

When, through the deliberate murder of James King, it became evident that citizens were compelled to defend themselves even against officials; and that those whose duty it was to execute justice, must be brought to justice; these men calmly considered the situation; and though people were suspicious of each other, the movement proceeded with such care that the best elements of society, not only in San Francisco, but in other parts of the state were united; organized as a Vigilance committee, with plan of work, and necessary preparation.

When all was ready, many men, from other parts of the state, quietly entered San Francisco, and at a preconcerted moment most of the city officials were arrested; the armory opened, and arms distributed to the already organized citizen soldiers; redoubts built of sacks filled with sand; cannon mounted, trusty guards placed on duty as police; and finally courts were convened and the offenders placed on trial.

Of course the Governor called out the state militia; but it not being well organized, those who were interested generally joined the "Vigilants," leaving him powerless. For once there was "government of the people, for the people and by the people."

Meantime the improvised courts completed their work. Some of the prisoners were executed; some allowed to leave the state under promise never to return; and certain officials, against whom there was only suspicion, were permitted to resume office.

When the work was all completed, the sand bag forts and cannon were removed, the state arms put in good condition, and restored to the armory, all office keys delivered to the proper officials, and then the committees adjourned and retired to their respective homes.

Should it be necessary, the citizens, in their organized strength, could promptly reassemble; but their work had been so well done there was no such need, and for years after that outburst of popular indignation, perhaps no city was as well governed as San Francisco.

Returning homeward, it was my intention to land at Acapulco, on the west coast of Mexico, and, taking the national road, visit the capital; thence to some port on the gulf; whence, after seeing something of that very interesting country and its people, I could ship for New York. But the day we sailed from San Francisco, the morning papers announced that Gen. Santa Anna had dissolved the Mexican congress by military force, and that the country was in the throes of revolution.

This made me regret my plan, especially when I found a Mexican among the passengers who said he had personal letters confirming the report. He was bitterly opposed to Santa Anna, gave many instances of his treachery to the Mexican government, expressed the opinion that he was plotting the downfall of the republic, and the establishment of a monarchy upon its ruins.

The cabin and deck of an ocean steamer afford unusual facilities for people to become acquainted. I soon discovered that there were on board men from many interesting parts of the world; and by approaching such in a quiet questioning way, in some secluded corner, much that was instructive and entertaining might be learned.

Among the passengers was an Irishman named O'Donohugh, who, with Thomas F. Meagher and others, having rebelled against the English government, were tried for treason, and banished to Van Dieman's Land; but making his escape, reached San Francisco in an American vessel, and was now on his way to New York.

Monday, July 4th, was celebrated at sea, and O'Donohugh being orator of the day, portrayed the perfidy of England, the wrongs of Ireland, the rise and fall of the rebellion; his trial, banishment and escape to a ship flying the United States' flag, which, he asserted, was the only flag in the world that could give him protection.

A splendid dinner was served, and I noticed that many of the passengers drank to excess; and there was revelry and confusion in the cabins. Retiring early to my state room to avoid the noise and drunken men, although the sea was calm, hours passed before slumber came to my relief.

Shortly after midnight, aroused by the intense heat of my room, I hastily dressed and sought the deck. Most of the revelers had become quiet; but as I looked out on the quarter deck it was evident that the officer in command was too drunk to attend to any duty.

Attracted to the door of the engine room by the boisterous mirth of the engineer and his assistant, the view was not assuring. The place overheated, and the machinery evidently laboring under a tremendous strain.

On the stairway leading down to the furnaees, stood a man who seemed to be in command of the coal heavers, shouting occasionally, "Shove her up there. Don't let the fire go out. Turn on the draft" &c.

The heat was intense, and the men, clearly under the inspiration of strong drink, were doing their best, but there was something frightful in the red glare of the fires; and the ponderous machinery in rapid motion, made the great ship, over three hundred feet long, tremble from end to end.

Slowly ascending to the upper deck, pausing a moment at the door of my room to find its heat insufferable, I found a company of passengers, who, like myself, had been driven from their rooms by the heat which pervaded the central part of the ship.

We were somewhere west or south of cape San Lucas, and had plenty of sea room, and as for accidents, boiler explosion, breakage of machinery, or the ship taking fire, whatever might happen; we could only do the best possible according to events and conditions. But I would have felt much safer on land, or on a ship controlled by sane men. However, laying upon a bench, under the star-gemmed sky, I slept uneasily until morning.

Grateful that we had been preserved from accident, glad the celebration was over, and that a sober crew was again in charge of the ship, I sought my state room, now cooled off, and very comfortable, and during the remainder of the voyage experienced no inconvenience from heat.

Reaching Acapulco, where our vessel

stopped a short time, I learned from the American consul that a division of Santa Anna's army was encamped near the city; that all coaches had been removed from the national road, and that private travel, even if possible, would be extremely dangerous. So, returning to the steamer, I concluded to go by way of Central America.

Nearing the west coast of Nicaragua, between two mountain-like promontories, our steamer entered a small bay at the head of which was the town of *San Juan del Sur* (St. John of the South.) There was no wharf, the ship anchored, and the passengers and baggage were sent part way in boats, and were met by natives, who, wading through the water, carried them on their shoulders to dry ground.

Seated on the brawny shoulders of a Zambo, (a person of Indian and Negro blood,) while another carried my satchel, I was borne high and dry through the surf.

I inquired, "How much do you charge?"

"One dime each."

My smallest change was a Five Franc silver piece, current at ninety-five cents. Handing it to one I remarked, "You must both take your pay out of that."

Without thought of seeing either of them again, I went some distance to a corral to obtain a mule, and just before starting for Virgin Bay on lake Nicaragua, one of the men brought me the exact change; fifty cents in coin, two pieces of soap, at ten cents each, and a cake of chocolate at five cents.

I mention this as an evidence of the honesty of these people, and also to call attention to the fact that articles of merchandise, rather than coins were used as money. Buying two cents worth of bananas, I gave the five cent cake of chocolate, receiving in change three rows of pins, at one cent a row; these current articles being always accepted without complaint. The only inconvenience, you needed a basket instead of a purse in which to carry your change.

From San Juan del Sur to Virgin bay was twelve miles. First ascending a steep mountain ridge, and then by a gradual descent to the lake, where we arrived late in the afternoon.

The expected boats to take us down the lake, had not arrived, and the principal hotel being overcrowded, several of us were directed to another, where, in broken English, we were very cordially welcomed by the proprietor, an elderly Spaniard.

While making a hurried tour through the village, night came with scarcely a warning of twilight. However, a weird light reflected from the mists around the crest of a volcano on the island of Ometepe, in the lake, enabled me to find the way to my hotel.

Now the town seemed more populous than by day. Lights were hung out, and the verandas were thronged with merry, laughing, talking, singing groups. At one place under an archway of bamboo, supporting a roof of palm leaves, a large company of ladies and gentlemen were engaged in a dance. As it was near the street, I paused a few minutes, listening to the lively music of the guitar and tambourine, and watching the strange yet graceful movements of the dancers.

The gentlemen wore a row of metalic buttons on the outside of their pantaloons from the knee down, and the usual buttoned and braided Spanish jacket. The ladies with dark skirt and bright colored basque, some with a slight, turban-shaped band around the head, but most of them bareheaded; their black hair tastefully arranged in heavy braids, and studded with what seemed the most brilliant gems.

Returning to my hotel I told the landlord about the dance, and the profuse display of gems. With a dignified smile he replied,

"They are not real gems, but lantern

flies, which the ladies at night pin in their hair."

"But is not that very cruel?"

"O, no," said he, "they do not stick the pin into the fly; only fix it so as to bind the fly in place, and when set at liberty it flies away unhurt.

## CHAPTER XXV.

*Chicken or Monkey.—Volcano of Ometepe.—Tost from my Hammock.—"Nothing but an Earthquake."—The Lighter.—Boy Caught by an Alligator.—Island of Ometepe.—Earthquake Injures Navigation.—San Carlos.—Ruined Fort.—Soldiers.—Down the River San Juan.—Castillo.—Impenetrable Forest.—Costa Rica.—Greytown.—On Board the Northern Light.—Greytown Bombarded and Burnt.*

Our hotel furnished no printed "bill of fare," but at supper our host came into the dining room and politely announced, "Bread, yams, chicken, eggs," and various other things difficult to remember, closing with, "tea, coffee and chocolate; all or part as you have choice."

I could not remember having tasted chicken since entering California, and very naturally had an appetite for it. But when brought, it had little to remind me of chicken,—perhaps I had forgotten,—slightly resembling certain parts of wild fowl, but this had been skinned, not picked.

Still, trying to make allowance for latitude and cooking, I almost persuaded myself that it was delicious chicken, until an attendant, who spoke only Spanish, entered, and I inquired of him, *Se llame pollo esto?* (Do you call this chicken?)

*No, Senor, se llama mono, esta bien, no?* (No sir, it is called monkey, very good, is it not?)

Coming over the ridge from San Juan del Sur, I had seen them by the road side, swinging on the branches of trees and chattering as we passed; but the idea of having one served for supper never entered my mind, and certainly failed to give relish to my evening meal.

The hotel was a bamboo structure rofed with palm leaves, and partitioned with the same material. My bed room was quite small, with a hammock suspended diagonally, about three feet from the floor.

Unaccustomed to such a bed, I found some difficulty in keeping properly balanced; and instead of the regular throb of the machinery, to which I had become used on board the ship, I could distinctly feel the tremble of the ground, and hear the rumble and boom of the volcano on an island in the lake; while the vapor above the mountain, probably reflecting the fires of the crater, looked like an immense flame.

Falling into a sound sleep, toward morning I waked to find myself on the matted floor. The room was dark; everything seemed in motion; and mingled with the thunders of the volcano came the roar of waters, as the surf is rolled upon the rocks by the incoming tide. It seemed as though lake Nicaragua was overwhelming the town.

Seeing a light in the office, bewildered, frightened, and but half dressed, I peered in. There sat the proprietor, holding a lighted candle in his hand, and smoking a cigarette.

I exclaimed, *Senor que es esta?* (Sir, what is this?)

Calmly he replied, "*Nada sino tierra temblor.*" (Nothing but an earthquake.)

His calmness allayed my fears, and entering into conversation, learned that for several days the volcano had been unusually active, and only a few minutes before, there was a severe earthquake shock; and I suppose it was this that caused me to fall from my hammock.

Daylight revealed the lake in great agitation, and although it seemed at its usual level, we could see where a wave had swept the lower part of the town, doing some damage to the bamboo buildings, though I think no lives were lost.

During the forenoon a steamboat came up the lake from San Carlos, but as there was no proper wharf, it could not approach the shore, and the only way of getting on board was by means of a "lighter;" a boat made of iron plates riveted together, like a steam boiler, with hollow, water-tight sides, which prevented it from capsizing or sinking; and there was not much danger of its breaking against the rocks.

This was brought to a rocky point, and as the wave carried it up nearly level with the top of the rock, those who were ready jumped in, and the receding wave carried the boat out into the lake, it was rowed to the steamboat, where leaving its passengers, returned for another load.

While this was in operation, a deeply affecting incident occurred. At another point about twenty rods from the boat-landing, a boy, said to be seven or eight years old, a native of the town, walked down the shelving rocks to a point washed by the waves. An alligator from some place nearby, came suddenly upon him, cutting off his retreat. With frantic gestures and screams he appealed for help, but before assistance could be given, was seized and both boy and alligator disappeared beneath the waters of the lake.

These reptiles, called by the natives *Cayman*, are very numerous in this country, some growing to a length of eighteen feet, and are found along the lakes and streams; often lying on the bank of the river or lake, their bodies accommodated to the curve of the bank, and their head resting on a level with the water. Their strong jaws, red inside, and adorned with rows of white teeth, seemed to open and shut like a pair of shears.

When the passengers were all on board, our little boat steamed out near the island of Ometepe; said to be about twenty miles long, north and south, and about seven wide. The volcano is rather north of the center, but the whole island was so enveloped in mist or steam, in many places descending to the water, that it resembled a vast cloud resting upon the lake; from the center of which came the heavy rumble of the volcano, like the roar of a distant thunder storm.

In the afternoon we reached the outlet of the lake, and source of the San Juan river which flows into the Caribbean sea. Here the recent earthquake, by raising a bar across the outlet of the lake, not only prevented our boat from going further, but by shutting off the usual flow of water, for a time greatly interfered with the navigation of the river.

This however, enabled me to visit the old town and fort of San Carlos. The village occupied a beautiful highland, overlooking both lake and river; but the houses were the usual low, bamboo, palm-leaf covered style common to this country.

The fort, once commanding the lake shore and outlet, built of stone, with a moat about ten feet wide, was a mass of ruins. To all appearance the wreck had been wrought by an explosion many years ago. Many beautiful brass or bronze cannon were mixed in, or underlying the demolished walls.

Near the ruined fort was a long shed used as barracks by a company of soldiers. Clothed in pantaloons and jacket of unbleached sheeting, barefoot, but wearing a small Panama hat, and armed with heavy Austrian muskets, their appearance was rather grotesque than military. However, what was lacking in the dress and uniform of subalterns and privates, was amply made up by the commissioned officers, who were brilliant in bright colors, gold lace, and polished leather. And probably there was the same contrast in their pay.

After a delay of about twenty-four hours, two small scow-like boats were brought up, to which we transferred, and floated down to the head of Castillo rapids. Going around these on foot, we stopped for the night at the little hamlet of Castillo.

At this transfer, a highly prized gift

from my old Indian friend Capitan Juan, which, neatly boxed, I had kept with special care, was stolen. See chapter XVI.

While coming down the river, I greatly admired the forest scenery on either bank, and was glad to get ashore, anticipating a stroll among the beautiful trees; but it was a case where

"Distance lends enchantment to the view."

It was impossible to take a single step from the beaten path. Prickly vines were so closely interlaced that it would be necessary to cut them at the ground, overhead, on both sides, and even then, the great, green, thorny vines would bar your way in front.

The next morning, on a small steam boat, the journey was continued down the San Juan. Just below the mouth of the San Carlos, a large tributary from the south, a short stop was made at Ochoa, in the republic of Costa Rica, and amid swarms of alligators, and the most beautiful and wonderful vegetation, with enough rain every day to keep it fresh and clean; at last, passing the low corals at the mouth of the river, our little boat began to rise and fall, on the swells that rolled in from the Caribbean sea; gliding safely into the bay of San Juan del Norte, or Greytown.

The steamship Northern Light lay at anchor in the bay, we were taken on board, and soon settled cozy enough on this floating palace. It was not so large as the Sierra Nevada, which brought us from San Francisco, but in form and finish, without and within seemed absolutely perfect.

It was late in the afternoon when we came on board, and I was greatly pleased at being permitted to accompany an officer of the ship to town. However our stay was too brief to obtain more than a glimpse of one of the main streets. There were some quite large buildings, but most of the houses were of the usual pattern seen in this country—bamboo, thatched with palm leaves.

After supper, with several of the passengers, I asked the privilege of again being taken ashore, but the officer explained that it would not be well for us to go in the evening; if however, a boat load of passengers, expected from the Pacific coast, did not arrive during the night, there would be an opportunity for us to go in the morning.

He said there was some misunderstanding between the republic of Nicaragua and the United States. A man-of-war had been sent to protect our interests; and soon after a British war ship arrived, perhaps to see fair play, but more probably for the reason that at that time, the neighboring so called Mosquito Reserve was under the protectorate of England; and it was not until 1894 that it was reincorporated with the republic of Nicaragua.

Of course the Passenger Line was careful not to give any occasion for trouble.

Sometime after this Senator Borland was sent by the United States' government to adjust matters, but a misunderstanding arose between him and the governor of Nicaragua, which induced Borland to take shelter on the American man-of-war. From this he endeavored to renew negotiations, but the governor was now on his dignity, and refused further parley. All efforts failed to elicit a reply.

At last Lieut. Ingrahm, commanding the man-of-war, sent a formal demand for a reply within a specified time, under penalty of firing on the town.

There was a fort on a point north of Greytown; but a vessel near enough to bombard the town from the south would probably be out of the range of its guns.

As the time approached Lieut. Ingrahm brought his ship into position, and signaled his readiness to begin. And when the time expired without response, he opened fire, demolishing the houses within his range, and then, sending a boat's crew ashore, burned the ruins.

The inhabitants had ample warning, and probably few, if any, were killed; but their property was destroyed without accomplishing any good. It was a cowardly act of war, and surely a discredit to the nation.

## CHAPTER XXVI.

*Delightful Outlook.—Refreshing Sleep.—The Caribbean Sea.—Water Spouts.—Struck by Lightning.—Ship on fire and in a Hurricane.—Passengers Frightened.—Fire Quenched.—Fair Weather—Reach New York.—Over the Catskill Mountains.—Visit Uncle Robinson's Family.—Trip to Fergusonville Academy—Again the Unexpected Happens.—Sad Hearts.—At Walton.—Meet Brother Edward.—Arrive Safely in Wisconsin.—Conclusion.*

### CONCLUSION.

The window of my state room looked out upon the town, and the wide array of lights made it appear larger by night than by day, while the phosphorescent flash of the tropical waves rolling in the distance resembled a sea of fire.

Wearied with the day's journey, under the influence of the weird lights and soft, night air, I fell into a profound sleep, and did not wake until sunrise. Noticing the tremble of the ship, and the regular throb of the machinery, I looked from my window, and instead of the town, saw only the wide expanse of ocean.

It was difficult to realize how deep my slumber had been. The expected passengers had arrived, anchor had been weighed, the ponderous engines started, and the parting signal fired withont even disturbing my rest.

Hastily dressing and going on deck, the merest trace of land lay off in the direction of the Mosquito coast, and on all other sides the horizontal line was an equal blending of sea and sky.

Sea voyages, and life on ship board, in their daily routine, though interesting, have been described so often, that I leave out all but a few special items.

After leaving Greytown, for several days the sea was very rough, with an occasional squall and dash of rain. While on deck one afternoon, three waterspouts appeared, one quite near, and coming toward the ship. A dark mass of cloud arched us like a frowning cliff, and the column of water which connected it with the sea, seemed running downward rather than upward, as in pictures they had appeared to my boyish fancy.

Presently all passengers were ordered below, and the hatches battened down. Just as I started down the companion way, lightning rent the main mast to splinters, and sent the rigging, spars and pulleys flying in the wind across the deck, Several men in the gangway were prostrated by the electric shock; the two engineers sprang from the engine room, which, as I looked inside through the open door, seemed a mass of flame.

The door was instantly closed, but the passengers raised the cry, "Fire! fire! the ship is on fire!"

Some fainted, many became wild with terror. Screams, prayers, sobs, and even curses, emphasized the confusion. Standing on the lower steps of the companion way, like one in a dream, I surveyed the scene.

The motion of the ship, the roar of the floods, the waters trickling through every seam, indicated that we were in the grip of the hurricane; and, considering the storm without and the fire within, escape for awhile seemed hopeless; and, with the rest, I believed the ship would be lost.

However, the crew, under the direction of competent officers, was at work; and in a few minutes men were ready with a large hose attached to a steam pump to put out the flames. The engine room door was opened; dense volumes of smoke rolled out, but no fire was found within. The closed doors, and water filled seams had smothered it; but along one side the dry, oil soaked wood, charred and blackened, showed how intense the fire had been.

In about two hours the hurricane was over, the wreckage of the mast had been cleared away, and with feelings of safety, and of gratitude to God, we were permitted again to pace the deck.

A rough sea and an occasional squall broke the monotony of the voyage, as we sailed out of the Caribbean sea, and across a corner of the gulf of Mexico. Stopping a short time at Havana, our good ship in due time entered the Gulf Stream, probably the greatest river in the world, with its warm water current and cold water banks; encountered the usual gale off cape Hattras, and one beautiful, calm, moonlight night sailed into the harbor of New York.

Two days afterward, crossing the Catskill mountains, I visited the family of my uncle, B. H. Robinson at Prattsville; and a week later started to visit several cousins of my father, who owned and conducted a boarding academy, in a country place, named after the proprietors, Fergusonville.

By stage from Prattsville, I reached the nearest village, East Davenport, about 7 P. M., put up at the hotel, intending to visit Fergusonville, three miles distant, the next morning.

While supper was in preparation the landlord informed me that there would be some delay, as more than twenty girls from the school had arrived and requested supper.

Contrary to the rules, they were out on a lark; had come on foot, part way through the woods, expecting to be back before they were missed. With the landlord and his wife, I joined them at the table; and for awhile they were certainly the wildest, giddiest, merriest company I ever saw.

When supper was nearly over, one ran to the door, and returning exclaimed,

"O, girls! it is raining!" and a dozen voices answered, "O, what shall we do?" And the light and music, and cheer faded out of some twenty happy hearts.

It rained all night, and the next morning the landlord provided conveyances, and the penitent company returned to school. While their enterprise amounted to much more than they anticipated, still, according to their expectations, it was not a success.

Visiting the school I spent the day with Rev. Samuel D. and Sanford I. Ferguson, Warden and Principal of the school, and their families. The school officials were solving the perplexing problem: how most kindly to deal with so many guilty of such a grievous disregard of discipline, and yet maintain the dignity of the school. The mystery of school discipline, like all other mysteries, is manifest in its results. So far as I could judge, every person connected with the school was as busy and happy as though nothing unusual had happened.

The campus embraced about three hundred acres, and the pupils were nearly all from large cities. In coming or returning, they were always accompanied by some responsible person. Arriving at the school, they enjoyed greater liberty than would be possible in village or city.

Outdoor exercise, the country air, life and influence of the Christian home; and the absence of corrupting allurements, made it a desirable place for young people. And many who lived in cities were willing to pay almost any price that their children might enjoy the benefits of that school.

At Walton village I met with brother Edward, and after a few weeks among the friends and haunts of our early boyhood, returned to Wisconsin, reaching Lodi in the early days of September, 1853, after an absence of about three and a half years. Years of some financial profit; not wasted as regards intellectual improvement; of considerable value in personal experience; and more than ever anxious to take up my suspended studies, and, if possible, realize my cherished dream of becoming thoroughly fitted for the work of an educator.

But the boy who penned the preceding pages, had just passed his twenty-first year; boyhood had given place to manhood; *In Camp and Cabin* is part of the record of his lost youth, and his feelings found expression in the words of Longfellow's minor strain: "MY LOST YOUTH."

And thus,

"my youth comes back to me,
And a verse of a Lapland song
Is haunting my memory still:
'A boy's will is the wind's will,
And the thoughts of youth are long, long
thoughts.'

There are things of which I may not speak;
There are dreams that cannot die;
There are thoughts that make the strong heart
weak,
Aud bring a pallor into the cheek,
And a mist before the eye
And the words of that fatal song
Come over me like a chill:
'A boy's will is the wind's will,
And the thoughts of youth are long, long
thoughts.' "

www.ingramcontent.com/pod-product-compliance
Lightning Source LLC
Chambersburg PA
CBHW081741220526
45468CB00008B/2196
* 9 7 8 1 7 2 0 7 9 6 5 8 9 *